GEOLOGY
FROM
EXPERIENCE

GEOLOGY FROM EXPERIENCE

Hands-On Labs and Problems in Physical Geology

With Enrichment Materials at
www.whfreeman.com/petersdavis

E. Kirsten Peters
Washington State University
Pullman, Washington

Larry E. Davis
Saint John's University
Collegeville, Minnesota

W. H. Freeman and Company
New York

Acquisitions Editor: Nicole Folchetti
Development Editor: Fred Schroyer
Project Editor: Jane O'Neill
Text and Cover Designer: Cambraia F. Fernandes
Illustration Coordinators: Bill Page, Cecilia Varas
Text Illustrations: Network Graphics
Photo Researcher: Jennifer MacMillan
Production Coordinator: Paul W. Rohloff
Composition: PRD Group
Manufacturing: R. R. Donnelley
Marketing Director: John Britch

Credit: Maps in back pocket of book (geologic map of the Pine Grove, Pennsylvania, Quadrangle, and topographic maps of Clark Fork, Idaho, and Palmyra, New York) courtesy of U.S. Geological Survey.

Library of Congress Cataloging-in-Publication Data

Peters, E. K.
    Geology from experience : hands-on labs and problems in physical geology / E. Kirsten Peters, Larry E. Davis.
        p. cm.
    "With enrichment material at www.whfreeman.com/petersdavis."
    Includes index.
    ISBN 0-7167-3145-2
    1. Physical geology—Laboratory manuals. I. Davis, Larry E. (Larry Eugene) II. Title.

QE44 .P47 2000
550′.78—dc21   00-060993

Printed in the United States of America
Second printing, 2003

W. H. Freeman and Company
41 Madison Avenue, New York, NY 10010
Houndmills, Basingstoke RG21 6XS, England

CONTENTS IN BRIEF

CONTENTS

E. Kirsten Peters studied at Princeton University and graduated summa cum laude with a degree in geology. She earned her Ph.D. from the Earth and Planetary Sciences Department at Harvard University. Her training is in geochemistry, petrology, and economic geology. She has interests in the history of geology as a discipline and in the tensions between science and religion. Her technical articles have appeared in *Geochimica et Cosmochimica Acta, The Journal of Economic Geology,* and *The Journal of Geological Education.*

Dr. Peters has written *No Stone Unturned,* a nontraditional introductory geology textbook emphasizing the history and philosophy of science, published by W. H. Freeman and Company. Under a pen name, she has also written a series of critically acclaimed mysteries, published by Random House and St. Martin's Press. Dr. Peters teaches geology at Washington State University in her hometown of Pullman, Washington.

Larry E. Davis received his Ph.D. in geology from Washington State University. His training includes marine biology, biostratigraphy, and sedimentology. He has published technical articles in *Lethaia, The Journal of Paleontology, The U.S. Geological Survey Bulletin,* and *The Journal of Geological Education.*

Dr. Davis's teaching honors include the Dean's Award for Excellence in Science Teaching, the President's Faculty Excellence Award for Instruction, the Washington Science Teacher of the Year in Higher Education Award, and the Society of College Science Teachers' Outstanding Undergraduate Science Teacher Award. He now teaches geology at Saint John's University in Collegeville, Minnesota.

To the Instructor (and the Curious Student)

An Attempt at Reformation

In our view, descriptive vocabulary has all too often become the focus of teaching geology and the Earth sciences at the introductory level. We like to *think* that vocabulary should not dominate what we teach—especially what we teach in lab to nonscience majors. After all, we want to teach something of the *spirit* of natural science. Nevertheless, textbooks and laboratory manuals in today's market ask students to learn hundreds of boldface, polysyllabic terms—porphyritic, pyroclastic, phaneritic, phenocryst, debris avalanche, debris fall, debris flow, debris slide—and on and on. Worse, they have few or no problems to be solved by the student, and they lack experimental labs.

In writing *Geology from Experience,* we have created a lab manual that is a significant departure from the norm.

Features

- **Hands-on and quantitative experiments** to be done in laboratory meetings, generally in small groups. These experiments have been class-tested by hundreds of students. They reflect our very hands-on approach and favor inquiry-based experimentation rather than traditional observation. By engaging students in active lab work, *Geology from Experience* emphasizes learning processes and ideas over memorizing terminology. The book's second unit presents lab exercises that set up this emphasis by reviewing both basic lab skills and specifically geological lab techniques. Students record their answers to all lab questions on the Lab Answer Sheets (perforated for grading) at the end of each unit.

- **Scientific problem solving,** which can be done by individual students in connection with lecture material, or singly or in groups in a lab or discussion meeting. A goal of *Geology from Experience* is to address the lack of fundamental scientific background among students enrolled in the physical geology course. Students are asked to use geological tools. They analyze data from observation, experiment, and research. They solve simple equations, and they make assessments and relevant predictions. The first unit includes lab exercises that review basic quantitative and qualitative tools that students should possess before taking this course or any science course. The problems may be done either at home or in lab. Students record their answers on the Problem Answer Sheets (perforated for grading) at the end of each unit.

- **GeoDetectives at Work.** Many units include fascinating stories that highlight the practical applications of geologic concepts by showing how geologists have applied their knowledge to help the FBI and local police with criminal investigations. Sand, soils, mineral grains, and even a metamorphic rock have been crucial to solving criminal cases in the United States and elsewhere. (Examples are traces of sediment in people's shoes, sand in the weights on World War II Japanese balloon bombs, distinctive rocks found far from their customary locales, etc.) Students read the stories and answer interpretive questions using their growing knowledge of geology.

Student Response to Our Approach

We have tried the exercises and problems in this book on hundreds of undergraduate students. You may be surprised to learn that we have not encountered resistance! Our A students, who adapt easily to whatever comes their way, respond very well. And our students who operate at lower grade levels find that the lab experiments and problem sets clarify the subjects they hear about in lectures. Our students seem to agree with us that they deserve a taste of the scientific experience rather than just hearing it discussed.

Organization

Unit 1 is a comprehensive review of skills needed to use this book. It can be assigned as homework in the first week of class, and Unit 2 can be used for the first laboratory meeting. Units 2 through 18 make up the core of this manual.

Units 12 and 13 cover topographic and geological maps. Each has a familiarization *prelab* that lets students explore real maps before plunging into the labs and problems. Two full-size, full-color topographic maps and one geologic map are provided in an envelope at the back of the book.

Supplements

An important feature of this book is supplemental material available at the *Geology from Experience* **Web site** : www.whfreeman.com/petersdavis

This site offers more problems and examples of basic empirical thinking in geologic situations and questions designed to lead students individually into library and Internet research. These examples demonstrate the relevance of geology and its sturdy connections with economics, history, politics, and engineering.

An **Instructor's Manual** is available for adopters. Because this lab manual is different from other commercial manuals, we developed our *Instructor's Manual* concurrently with this text to guide both veteran and less-experienced instructors. The *Instructor's Manual* describes the labs, explains what materials are needed and how to acquire and set them up, and offers advice about commercially available materials.

Sources for Materials Used in Our Labs

The diverse materials we use are available from grocery, drug, and building-supply stores; supply sources at your school (such as building maintenance and chemical lab supply); bookstores; Ward's Natural Science Establishment, Inc. (P.O. Box 92912, Rochester, NY 14692, 800-962-2660, fax 800-635-8439, or in Canada 800-387-7822, www.wardsci.com/); and Edmund Scientific (101 East Gloucester Pike, Barrington, NJ 08007-1380, 800-728-6999, fax 856-547-3292, www.edsci.com/).

Acknowledgments

It's a pleasure to acknowledge the many people who have helped us prepare this book. Most important, the late Sheldon Judson (Princeton University) inspired Kirsten Peters to attempt pedagogic reformation in freshman geology. Sheldon's encouragement was invaluable to this project, and the author still misses his advice and guidance every day.

The development editor, Fred Schroyer, has been a tireless and cheerful addition to the writing team. His help has been crucial to the completion of this book. Without him, we would surely have given up.

In addition, we want to thank several staff members at W. H. Freeman and Company for their help in making this book a reality. Their accomplishments were amazing, particularly given the tight deadlines imposed on the last stages of this project. Acquisitions editor Nicole Folchetti helped keep everyone focused on the deadlines for publication with her encouragement and enthusiasm. Project editor Jane O'Neill kept the production process on track and the authors in the loop, which was greatly appreciated. Editorial assistant Bradley Umbaugh was invaluable in his help in preparing the manuscript for production and in solving the many crises that cropped up along the way. And free-lance copy editor Penny Hull contributed her excellent editorial skills to the process.

Legions of students here at Washington State University have undertaken the laboratory projects and problems in this book. Their feedback helped improve the content presented here, and we are enormously grateful to them. Numerous reviewers helped us further refine the book. They include Gary C. Allen, *University of New Orleans;* Allen W. Archer, *Kansas State University;* David M. Best, *Northern Arizona University;* A. L. Bloom, *Cornell University;* Stephen K. Boss, *University of Arkansas;* Roderic Brame, *Virginia Tech;* H. Robert Burger, *Smith College;* Ernest H. Carlson, *Kent State University;* Roseann J. Carlson, *Tidewater Community College;* James R. Carr, *University of Nevada–Reno;* George S. Clark, *University of Manitoba;* Peter Copeland, *University of Houston;* James M. Cronoble, *Metropolitan State College of Denver;* Cameron Davidson, *Beloit College;* Carl N. Drummond, *Indiana University–Purdue University at Fort Wayne;* Nels Forsman, *University of North Dakota;* Andrew Frank, *Chemeketa Community College;* Fred Goldstein, *College of New Jersey;* L. M. Gray, *Mount Union College;* Warren D. Huff, *University of Cincinnati;* David T. King, *Auburn University;* Hobart M. King, *Mansfield University of Pennsylvania;* Albert M. Kudo, *University of New Mexico;* Norman P. Lasca, *University of Wisconsin–Milwaukee;* Michael Manga, *University of Oregon;* G. David Mattison, *Butte Community College;* D. Brooks McKinney, *Hobart and William Smith Colleges;* Peter Muller, *State University of New York–Oneonta;* Rainer Newberry, *University of Alaska;* Jonathan A. Nourse, *California State Polytechnic University;* J. K. Osmond, *Florida State University;* James S. Reichard, *Georgia Southern University;* John J. Renton, *West Virginia University;* Rob Sternberg, *Franklin and Marshall College;* Christopher A. Suczek, *Western Washington University;* and Dr. Andrew H. Wulff, *University of Iowa.*

WEB SITE DIRECTORY

The chart below lists the enrichment material available on the Web site, which includes practice problems and research projects.

Unit	Practice Problems	Research Projects
1	W1.1. Converting Kilograms to Troy Ounces W1.2. Unit Analysis of Formula for Volume of a Sphere W1.3. Normal Sand/Gravel Distribution (%) W1.4. The Le Châtelier Principle and Dissolving an Unknown Substance on Another Planet W1.5. Expressing Speed of Sound in Feet per Minute W1.6. Expressing Speed of Sound in Fathoms per Hour W1.7. Unit Analysis of the Formula for Movement of a Particle in a Straight Line under Constant Acceleration W1.8. Unit Analysis of Formula for Volume of a Cylinder W1.9. General Shapes of Common Histograms	NA
2	NA	NA
3	W3.1. Identifying Minerals by Mineral Mass Color versus Streak Color W3.2. Common Effervescent Minerals	Why Diamonds Are DeBeer's Best Friend
4	W4.1. Plutonic Equivalents of Surface Igneous Rocks W4.2. Identifying Plutonic Rocks W4.3 How Composition Varies with Temperature	Mount St. Helens Blows Her Top
5	W5.1. Particle-Size Distribution in a Whitewater Streambed W5.2. Sediment Types along a Temperate Beach	The California Gold Rush of 1849
6	W6.1. Metamorphic Rock from a Rifle W6.2. Metamorphic Grade W6.3. Fracturing Quartz Sandstone versus Fracturing Quartzite W6.4. Metamorphic Terranes W6.5. Type of Metamorphism W6.6. Inferring Temperature and Pressure Conditions	Michelangelo and His Most Seductive Rocks
7	W7.1. Determining Relative Age W7.2. Principles of Geologic Dating	Haggis, Whiskey, Wordsworth, and Metamorphism

Unit	Practice Problems	Research Projects
8	W8.1. Determining Earth's Age from Changing Sea Levels W8.2. Determining Age from Sediment Thickness W8.3. Determining Age by Radioactive Decay W8.4. Determining Earth's Age from Changing Sea Levels W8.5. Determining Age from Sediment Thickness W8.6. Determining Age by Radioactive Decay W8.7. Exponential Change into the Future W8.8. Determining Age by Radioactive Decay	A Nuclear Life and Death
9	W9.1. Plotting Recent Earthquakes W9.2. The Hawaiian Islands	Alfred Wegener Rocks the Geologic Boat
10	W10.1. Saving Your Life in a Quake W10.2. Be Prepared for an Earthquake W10.3. Types of Seismic Waves W10.4. Earthquakes and Tsunamis	The Northridge, California, Quake
11	NA	Not One Tree-Hugger in the Crowd
12	W12.1. Where Are You? (Latitude and Longitude) W12.2. Where Are You? (Township and Range) W12.3. Magnetic Declination	NA
13	W13.1. Tectonic Strain Rate W13.2. Geology of the Island of Hawaii	NA
14	W14.1. Drawing a Stream Cross Section	Grand Forks and the Red River of the North
15	W15.1. Topographic Map of Thousand Springs, Idaho	Our Nation's Worst-Polluted Groundwater
16	W16.1. Meteorite Impact Simulation	Glacial Outburst Flooding
17	W17.1. Identifying Topographic Features Produced by Gradual Processes (Mount Spickard, Washington Quadrangle) W17.2. Composition of Sea Water	Gradual Change
18	W18.1. Profitability of a Gold Mine W18.2. Calculating Gold and Silver Production W18.3. Interpreting a Cross Section W18.4. The Largest Skarn in the World	The Klondike Gold Rush

Opening the Geologist's Tool Kit

Part I

Every profession has a specialized vocabulary. An accountant, for example, may speak of a "leveraged" deal on the stock market or declare that a "low-income family" qualifies for the government's "earned income credit." All three terms have specific, defined meanings. Similarly, an accountant uses specific tools to do her job. She may perform financial analysis with a balance sheet or computer spreadsheet. The point: to understand what any professional does, you must learn their *vocabulary* and understand their *tools*.

The same is true in science. That's why you will learn new vocabulary in this course and discover special tools for analyzing the world around us. It's easiest to learn the vocabulary and tools by doing small problems and experiments. Like learning to ride a bicycle, once you get the hang of it, you won't forget what you have mastered.

Part I of this lab manual contains Units 1 and 2, which summarize analytical methods. These methods are the basic tools of the trade in scientific work. You may have used some of them in high school, but revisiting them here will refresh your memory and improve your understanding. These units let you practice basic scientific skills before you jump into the specifics of geology.

Understanding the tools in this unit will lay a firm foundation for the rest of your geology course. *So please do all the work assigned in these units with great care, because you need to master these skills before continuing. Ask for help if you have difficulty!*

Paper-and-Pencil Tools   Unit 1

This unit presents paper-and-pencil tools used by geologists and other scientists. We offer examples and problems in unit conversion (like meters to feet), unit analysis of formulas, constructing graphs, and constructing histograms (bar graphs). We also offer problems in applying the Le Châtelier principle: a system at equilibrium responds to any change by working to minimize the change.

Tool 1.1   Unit Conversion

One of arithmetic's most useful tools is unit conversion–converting from inches to feet, liters to gallons, miles to kilometers, and so on. The key to unit conversion is that you can multiply any number by 1 without changing the value of the original number. For example,

$$5 \times 1 = 5$$
$$31 \times 1 = 31$$
$$\pi y^3 \times 1 = \pi y^3$$

Also, when you divide anything by itself, like $7 \div 7$, the result is 1. All such expressions are equal in value to the number 1. For example:

$$27 \div 27 = 1$$
$$\frac{168}{168} = 1$$
$$843/843 = 1$$

Further, you can multiply any number by a fraction like 1/1, abc/abc, or 62/62 without changing the value of the original number. This little trick allows us to change *units* to more useful forms without altering *value* in any way. As a first step toward converting units, we can write:

$$\frac{12 \text{ inches}}{1 \text{ foot}} = 1$$

This equality is true because the upper and lower numbers in the fraction have the same equivalent value expressed in different form. Similarly,

$$\frac{\$1}{100 \text{ cents}} = 1$$

In the preceding examples, the fractions contain mixed units in the numerator and denominator. Nevertheless, you know they equal 1, because the top and bottom are actually the same (that is, 12 inches = 1 foot and $1 = 100 cents).

This gives us an easy way to convert measurements. For example, if we want to convert a value expressed in feet into a value expressed in inches, we can easily do so and be right every time. For example:

How many inches are in 3.5 feet? We can write:

$$3.5 \text{ feet} \times \frac{12 \text{ inches}}{1 \text{ foot}} = ?$$

$$3.5 \text{ feet} \times \frac{12 \text{ inches}}{1 \text{ foot}} = ?$$

$$3.5 \times \frac{12 \text{ inches}}{1} = 42 \text{ inches}$$

That's the basic idea behind unit conversion. And, by the same reasoning, you can multiply a number by several fractions to achieve multiple conversions, as long as each fraction equals 1. This technique is simple and useful because it allows us to *manipulate* numbers, putting them in different form for convenience without really changing them. It's like pouring 12 ounces of beverage from a can into a glass: you change the shape of the drink, but it's still the same amount.

Example: Converting Square Feet to Square Yards. Suppose that you measure your studio apartment with a tape measure and find that it is 21 feet long and 15 feet wide. You want to buy carpet for the room and, armed with your measurements, you figure that you need

21 feet × 15 feet = 315 square feet of carpet
(also written as 315 ft.^{2})

At the store, however, you find that carpet is priced in square *yards*, not square feet. So, in the store's terms, how many square yards of carpet do you need? Your first thought might be to divide 315 by 3, but oops—wrong! Play it safe and write out a unit-conversion equation, converting square feet to square yards:

$$315 \text{ ft.}^2 = 315 \text{ feet} \times \text{feet} \times \frac{1 \text{ yard}}{3 \text{ feet}} \times \frac{1 \text{ yard}}{3 \text{ feet}} =$$

$$\frac{315 \text{ yards} \times \text{yard}}{3 \times 3} = 35 \text{ yd.}^2$$

Geologists often use unit conversion. The problems in this book assume that you can convert units using this method. You can convert units not only within the English system (like feet to yards or feet to inches) but from English units to metric ones (like feet to meters or gallons to liters) and vice versa.

The table on the inside front cover provides equivalent values for conversions. Each pair of numbers in this table, when written as a fraction, equals 1. Use these pairs to convert any units you encounter in this book.

Example: Converting Miles per Hour to Kilometers per Hour. Over-the-road truck drivers in the United States often cover 700 miles per day. Imagine that your cousin from England, where metric measures are used, moves to Louisiana and finds work as a truck driver. Using the method of unit conversion and the table on the inside front cover, help him to express 60 miles per hour, a common speed limit, in kilometers per hour.

$$\frac{60 \text{ miles}}{\text{hour}} \times \frac{1.609 \text{ km}}{\text{mile}} = \frac{96.5 \text{ km}}{\text{hour}}$$

Over a 14-hour workday, your cousin averages 52 miles per hour. How many kilometers has he traveled by the end of the day?

Start with what you know: 14 hours of travel. Then use cancellation of units to get the units you want: kilometers.

$$14 \text{ hours} \times \frac{52 \text{ miles}}{\text{hour}} \times \frac{1.609 \text{ kilometer}}{\text{mile}}$$
$$= 1171 \text{ kilometers}$$

Example: Converting Square Feet per Quart to Square Meters per Liter. Imagine that you have been working as a house painter for most of the summer. Your favorite brand of high-quality paint claims that it will cover 150 square feet per quart. One of your cousins, who sells paint in Canada, tells you that he can get you a better paint at the same price in U.S. dollars. "How much does a can cover?" you ask. He looks at his can's label and says, "11 square meters per liter." Which paint claims to cover more area, yours or his? Try converting the units on your brand.

$$\frac{150 \text{ foot} \times \text{foot}}{\text{quart}} \times \frac{0.3048 \text{ meter}}{\text{foot}} \times \frac{0.3048 \text{ meter}}{\text{foot}}$$
$$\times \frac{1.057 \text{ quart}}{\text{liter}} = 14.73 \text{ m}^2/\text{L}$$

Your paint appears to cover more area than your cousin's Canadian paint. Make sure that you see how the units cancel correctly—drawing lines through them is the best way, as shown—and make sure to rub in the answer at the next family reunion!

Problem 1.1. Volume of Spilled Oil. In 1989, an oil tanker in Prince William Sound, Alaska, hit a reef. In a matter of hours, about 10 million gallons of oil spilled into the water. This was the worst oil spill in U.S. history.

Q1.1. What volume of oil was released, expressed in liters? (Use the unit-conversion method taught in this book. Record your answers to all questions on the Answer Sheet at the end of this unit.)

Problem 1.2. Weight of Spilled Oil. Water weighs about 1 kilogram per liter. Ocean water is a bit heavier (denser) because of the salt dissolved in it. Oil, of course, is less dense than water, which is why it floats atop water, creating oil slicks and the rainbow effect you see in oily puddles along the street. Now, suppose that the crude oil spilled from the ruptured tanker in Alaska weighed 0.91 kg/L.

Q1.2. How many metric tons of oil were spilled (1 metric ton = 1000 kilograms)? (Use the unit-conversion method.)

Problem 1.3. Converting Metric Tons to Pounds and Cents to Dollars. Imagine that you are an international copper broker. Copper's price is 88 cents per pound. You have an overseas dealer who wants to sell 1488 metric tons of copper (1 metric ton = 1000 kilograms).

Q1.3. How much is the copper worth right now in U.S. dollars? (Use unit conversion.)

Problem 1.4. Converting Cubic Meters to Cubic Feet. Imagine that a geologist digs a hole where she is prospecting, removing 1 cubic meter of soil.

Q1.4. How much soil is that in cubic feet? (Use unit conversion.)

Problem 1.5. Converting Grams to Troy Ounces to Dollars. A geologist is hiking along a creek and discovers a nugget of pure gold. At her office, she weighs the nugget on the only scale available, which reads in grams. The nugget weighs 388 grams.

Q1.5. If gold is priced at $345 per troy ounce in U.S. dollars, what is the nugget worth in dollars today? (Use unit conversion: 12 troy oz. = 1 lb.)

Tool 1.2 Unit Analysis of Formulas

Unit analysis is an easy concept. It helps you think through the problem you are trying to solve and understand the units you are dealing with. When you write a formula, either from memory or from a reference, pause to check the units in the equation. The units must make sense for the formula to work! For example, consider this simple equation:

$$rate = \frac{distance}{time}$$

It makes sense as far as its units are concerned, for *rate* means a *distance traveled over time.*

To double-check, plug in some familiar units that you know well:

$$miles\ per\ hour\ [rate] = \frac{miles\ traveled\ [distance]}{hours\ of\ travel\ [time]}$$

You can see that this will work. This little trick of checking the units for reasonableness and plugging in familiar examples as a quick test will help you with all equations you encounter in all your classes—geology, math, chemistry, psychology, and so on. If you have any fear of math, this technique really helps!

Example: Unit Analysis of Formula: Area of a Circle. Imagine that you are taking a standardized test, perhaps for placement in the armed forces or admission to a professional school. This first problem asks you to find the area of a circle having a radius of *r*. You aren't quite sure you remember the formula for the area of a circle, but you make an attempt:

area = $2\pi r$
(area = 2 × pi × the radius of the circle)

Can this formula be correct? Substitute some units to see if your equation makes sense. For example:

area = 2 × π × length units

Or you might try specifics:

in.^{2} [area] = 2 × π × inches

In either case, you see that the formula cannot be correct because it equates (makes equal) a squared length unit (area) and a plain-and-simple length unit. In fact, what you remembered was the formula for the circumference of a circle. The formula for area is

$$\text{area} = \pi r^2$$

Here you can see that the units make sense, because the formula says that area—which will be a unit of length squared—equals a length measurement (radius) squared.

For the following problems, record your answers on the Answer Sheet at the end of the unit.

Problem 1.6. Unit Analysis of Formula for Area of a Triangle. Imagine that you are taking one of those long, tedious standardized exams. After studying one problem, you decide you need the formula to find the area of a triangle. You recall something like

$$\text{area} = \tfrac{1}{2} \text{ base} \times \text{height}$$

Q1.6. In terms of unit analysis, does the formula make sense?

Problem 1.7. Unit Analysis of Formula for Porosity. A sponge has many holes or spaces within it, a property called *porosity*. For geologists, porosity is a measure of exactly how much pore space is present in a rock or soil. You will encounter *porosity* in your geology course when you study *sandstones*, which usually are filled with tiny pores. You also will see an example of extreme porosity in *pumice*, a volcanic rock that has so many open spaces that it actually floats.

Suppose your geology teacher displays the formula for porosity:

$$\text{porosity} = \frac{\text{volume of pore space} \left(\text{cm}^3 \text{ or in.}^3\right) \text{ in a rock sample}}{\text{volume of rock sample} \left(\text{cm}^3 \text{ or in.}^3\right)}$$

Q1.7. What is the unit for porosity? (*Hint:* You may be surprised by the answer.)

Tool 1.3 Constructing Graphs

Graphs—scientists often call them **plots**—are useful for analyzing measurements of natural processes. The most common plot is the **x-y graph** (the horizontal axis is for *x* values; the vertical axis represents *y* values). You already are familiar with this concept, for *x-y* graphs are commonplace on TV and in newspapers and magazines.

As an example, the business pages of a newspaper might use an *x-y* plot to show the trend in wheat prices over several days (Figure 1.1).

In a more geological vein, consider an *x-y* graph of the *depth* of a river at a given time versus the river's *velocity* at that time. For example, let's use the depth and velocity of the Mississippi River at St. Louis (Figure 1.2). In this type of graph, you may see **error bars,** either generalized ones for all the data or bars that are assigned to specific data points. The point represents the measurement, and the error bar indicates the range of possible values for each variable. The figure indicates that the possible error involved in measuring the velocity of the river is large compared to the possible error in measuring river depth.

Note that the figure is not just a presentation of data points with their associated possible errors. The author of the graph also has used a statistical procedure to "fit" a curve through the data. Both curves and straight lines can be constructed for data using most business and scientific calculators.

Example: Exponential Equation, Logarithmic Graph, Semilogarithmic Graph. As it happens, in natural systems like our Earth and the life forms that inhabit it, many processes are governed by **exponential equations.** Such equations involve a variable that is raised to a power (exponent), like 2^x or e^x. Don't panic over such expressions. Just understand that these processes are often easiest to represent on a **logarithmic graph.**

You'll see that this is simple if we use a concrete example. Imagine that you live on a small, grassy island that has very few animal species. Someone from the mainland releases five pairs of rabbits on your island.

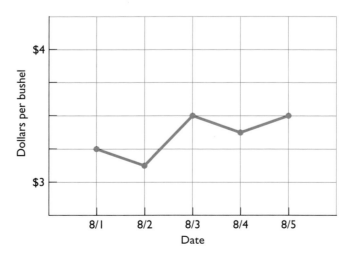

Figure 1.1 An x-y plot for wheat prices over several days.

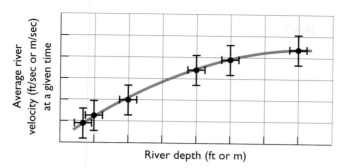

Figure 1.2   Graph (x-y plot) of river depth versus the river's average velocity at a certain point, for example, depth and velocity of the Mississippi River at St. Louis. The graph indicates that the deeper the river, the faster it flows. Note that the data do not lie in a straight line—they are not linear.

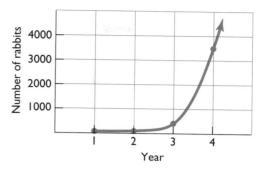

Figure 1.3   Unlimited rabbits: exponential curve. Unlimited reproduction of rabbits is represented by an exponential curve on a graph that has a linear horizontal scale and a linear vertical scale.

Because rabbits are not a native species on your island, there are no predators (coyotes, eagles, owls, and so on) to eat the rabbits. The rabbit population is free to increase at its natural rate, which is exponential.

You decide to count the rabbits and record their population. Let's assume that they all were released when young, that they will live at least four years, and that none die accidentally. If each pair of rabbits produced 12 babies per year, with the offspring evenly divided between male and female, your records would show year-by-year numbers like these:

Year 1: 5 pairs = 10 rabbits

Year 2: (5 pairs × 12 babies) + original 10 reproducing rabbits = 70 rabbits (35 pairs)

Year 3: (35 pairs × 12 babies) + 70 reproducing rabbits = 490 rabbits (245 pairs)

Year 4: (245 pairs × 12 babies) + 490 reproducing rabbits = 3430 rabbits

Summarizing:

Year 1: 10 rabbits

Year 2: 70 rabbits (7 times the original population)

Year 3: 490 rabbits (49 (or 7^2) times the original population)

Year 4: 3430 rabbits (343 (or 7^3) times the original population)

To graph this on a normal x-y plot, we must have a vertical (y) axis high enough to show 3430 (Figure 1.3).

(Note that no error bars are included in this graph because they aren't needed in this make-believe example.)

Obviously, as the rabbit population continues its exponential growth, the y axis of the graph must grow tremendously. Scientists find it more convenient to represent this kind of information on x-y graphs that have one logarithmic scale. This kind of plot is called **semilogarithmic.** The rabbit example is plotted semilogarithmically in Figure 1.4.

Example: Graphing Half-life of Carbon on Linear and Semilog Plots. Carbon-14 is a form of carbon used to date recent geologic events and even artifacts from human history. The method is accurate back to about 55,000 years ago. Carbon-14 is *radioactive* because the nucleus of a carbon-14 atom is unstable. Through time, each atom "decays" by giving off energy and becomes transformed into a different element.

Scientists have measured the rate at which carbon-14 decays. Remarkably, under all known conditions of

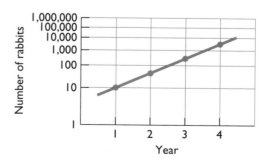

Figure 1.4   Unlimited rabbits: the same numbers on a semilogarithmic plot. Unlimited reproduction of rabbits is represented by a line on a graph that has a linear horizontal scale and a logarithmic vertical scale.

Table 1.1

How carbon-14 decays if we start with 1,000,000 atoms (half-life of carbon-14 is 5730 years)

Number of carbon-14 atoms	Time
1,000,000	0 (start)
500,000	5730 years later (1 half-life)
250,000	11,460 years after start (2 half-lives)
125,000	17,190 years after start (3 half-lives)
62,500	22,920 years after start (4 half-lives)
31,250	28,650 years after start (5 half-lives)
15,625	34,380 years after start (6 half-lives)
7,812	40,110 years after start (7 half-lives)
. . . and so on . . .	. . . and so on . . .

temperature and pressure, carbon-14 decays at a constant rate. One way of expressing this rate is to state the time required for half of a group of carbon-14 atoms to decay. This *half-life* of carbon is 5730 years. Thus, if a piece of charcoal contained 1 million atoms of carbon-14, half of them would be transformed into another element after the passage of 5730 years. From this information, we can construct Table 1.1.

Graphing these values on a regular grid gives the result shown in Figure 1.5. Because this illustrates a known principle, the graph includes no error bars. As you can imagine, for some purposes it can be useful to plot such numbers on a semilog graph (Figure 1.6).

Problem 1.8. Gold Nuggets—Size versus Silver Content. A geologist studies gold nuggets found by prospectors in Canada's Yukon Territory. He finds that each nugget contains at least some silver. He wonders if any systematic

relationship exists between nugget size and silver percentage. The easiest way to find out is to plot the two variables. You can do so using the geologist's analytical results, shown in Table 1.2.

Q1.8. In Figure 1.12 on your Answer Sheet, plot the nugget weight versus the silver percentage. Decide how to mark the x axis (the horizontal axis) to accommodate the data.

Q1.9. Looking at your plot, would you say that there is a relationship (even a rough one) between nugget size and proportion of silver in the gold?

Problem 1.9. Exponential Growth of Dalmation Dog Population. Dalmatian dogs (age 3 years), one male and one female, are released on a huge tropical island where they find plenty to eat and no animals prey on them or their offspring. If the female comes into heat twice a

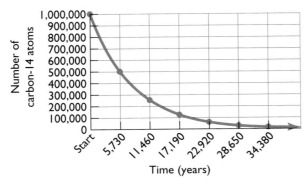

Figure 1.5 Half-life of carbon-14: linear graph. Both scales are linear.

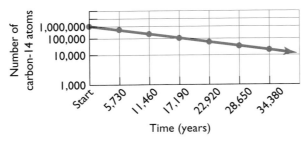

Figure 1.6 Half-life of carbon-14: semilogarithmic graph. The horizontal scale is linear; the vertical axis is a logarithmic scale.

Table 1.2

Gold nuggets found by prospectors in Canada's Yukon Territory

Nugget number	Weight of nugget (grams)	Percent of silver in the gold	Nugget number	Weight of nugget (grams)	Percent of silver in the gold
1	17.2	33	11	15.0	30
2	8.0	17	12	9.1	18
3	1.1	2	13	2.0	4
4	1.9	3	14	0.6	1
5	0.9	2	15	0.9	2
6	8.3	15	16	1.4	3
7	0.8	2	17	2.1	4
8	1.1	3	18	1.3	2
9	1.7	4	19	1.1	2
10	2.1	4	20	11.7	21

year and has 8 viable puppies in each litter, and if each pup is fertile by age 6 months, how many Dalmatians will there be on the island at the end of $2\frac{1}{2}$ years? (Assume that 50% of all puppies are female and that adult Dalmatians live for 12 years.) To help you with this problem, start by completing Table 1.3.

Q1.10. Plot your results on the normal (linear) grid in Figure 1.13 on your Answer Sheet. Determine how to mark the y axis to accommodate all the Dalmatians. When you have plotted your dots, connect them with your best hand-drawn curve.

Q1.11. Plot your results on the semilog graph paper in Figure 1.14 on your Answer Sheet. The plot has been started for you. When you have plotted all the dots, connect them with a line.

Q1.12. Describe the curve in Figure 1.13 in your own words.

Problem 1.10. Graphing the Decay of an Isotope (Focus on the Parent Element). Uranium-235 has a half-life of 704 million years. Imagine that a certain mineral (for example, zircon) contains 1024 atoms of uranium-235.

Table 1.3

Exponential growth of Dalmatian dog population

Time	Number of adult Dalmatians	Number of puppies	Total number of Dalmatians
Dogs released	2 (1 female)	0	2
$\frac{1}{2}$ year later	2 (1 female)	8 (4 females)	10
Full year later	10 (5 females)	40 (20 females)	50
$1\frac{1}{2}$ years later	50 (25 females)		
2 years later			
$2\frac{1}{2}$ years later			

Q1.13. On each of the two graphs in Figures 1.15 and 1.16 on your Answer Sheet, construct a curve to indicate the decreasing number of atoms of uranium-235 in the mineral over time. Both *x* and *y* axes have been clearly labeled for you on the linear graph paper (Figure 1.15) and the semilogarithmic graph paper (Figure 1.16).

Tool 1.4 Constructing Histograms

Another valuable graphical tool to geologists is the **histogram,** a type of plot that shows the *frequency distribution of a measurement.* You are probably familiar with histograms that show the distribution of class grades. In a class of 12 students, for example, the grade distribution might look like Figure 1.7.

From this histogram, you know instantly that more students (4 of them) earned a C than any other grade. One student flunked, two earned an A, and so on. Also, you can see that more students earned grades above C than below C. Thus, the distribution is not uniform but is skewed slightly. Still, for a small class, the grade distribution is close to a **normal curve** (also called a **bell curve** for its bell shape). An even more strongly skewed distribution of grades is shown in Figure 1.8.

Occasionally, a group of students ends up with quite a few high grades and quite a few low ones but few in the middle. Figure 1.9 is a histogram of such a **bimodal distribution,** that is, a distribution that has two distinct peaks.

Example: Histogram Showing Distribution of Sand Dune Grain Size. Geologists who study sand dunes use histograms to plot the distribution of sand grain sizes in a dune. Studying samples from one dune might yield a histogram that looks like Figure 1.10.

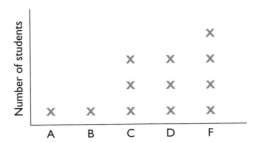

Figure 1.8 Histogram of skewed grade distribution among 12 students. Each X represents a student.

A geologist interpreting this histogram would reason as follows:

- The wind blew grains of sand around, perhaps blowing a lot of sand across many miles.
- Where this particular dune sits, the grains deposited measure mostly 1/8 to 1/16 inch.
- Smaller grains, being lighter, were carried onward to another location.
- Larger grains, being heavier, settled out sometime earlier at another location.

A histogram from a sand dune in another area might be quite different. By studying distributions of sand grain sizes, geologists have learned a great deal about how the wind moves particles. Work on many topics in other branches of geology has been aided by constructing histograms.

Problem 1.11. Grade Distribution in Percent. In a freshman geology class of 200 students, grades at semester's end are as shown in Table 1.4.

Q1.14. Convert the number of students who received each grade into a percentage of the whole group. Then draw a histogram for the grade

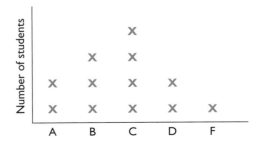

Figure 1.7 Histogram of grade distribution among 12 students. Each X represents a student.

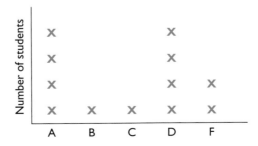

Figure 1.9 Histogram showing bimodal distribution of grades among 12 students. Each X represents a student.

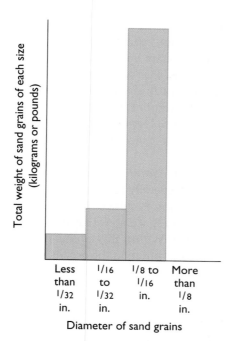

Total weight of sand grains of each size (kilograms or pounds)

| Less than ¹/32 in. | ¹/16 to ¹/32 in. | ¹/8 to ¹/16 in. | More than ¹/8 in. |

Diameter of sand grains

Figure 1.10 Histogram showing distribution of sand grain sizes in a sand sample from a dune, by weight.

distribution of the class on the grid in Figure 1.17 on your Answer Sheet.

Q1.15. What is the name for this type of distribution?

Problem 1.12. Gold in Parts per Million from 18 Samples in a Mine. Deep underground in a gold mine, a geologist takes 18 samples of rock. He sends the samples to an assay lab, and the lab returns the results shown in Table 1.5.

The term *ppm* means parts per million by weight. This means that for every weight unit of gold in the rock, there are 1 million weight units of rock (for example, 1 pound of gold in 1 million pounds of rock). The term ppm is common in the gold industry. Rocks that contain at least 1 ppm gold generally are worth mining. Another way to say this: the value of gold is high compared to lots of other Earth materials.

Q1.16. Using an **X** for each rock sample, construct a histogram of the geologist's results in Figure 1.18 on your Answer Sheet. If you find a value that has a second or third sample for it, simply stack the **X**s atop one another.

Table 1.4

Grade distribution for 200 students

A+ 4 students	B+ 6 students	C+ 10 students	D+ 30 students
A 24 students	B 5 students	C 18 students	D 10 students
A− 46 students	B− 5 students	C− 22 students	D− 15 students
			F 5 students

Table 1.5

Gold content in 18 samples from a mine

Rock sample	Gold content (ppm)	Rock sample	Gold content (ppm)	Rock sample	Gold content (ppm)
1	1.1	7	0.6	13	0.7
2	1.2	8	0.9	14	1.3
3	0.9	9	1.1	15	1.4
4	1.3	10	1.2	16	1.1
5	1.1	11	2.2	17	0.8
6	0.8	12	1.0	18	0.9

One rock sample has an unusually high gold concentration. In the science of statistics, such a misfit sample is called an **outlier,** because it lies outside the range of most other samples. Gold particles, due to their chemical nature, have a strong tendency to clump together, so such high-concentration outliers are not unusual in gold mining. This phenomenon is called the *nugget effect.*

Determining the average gold content of rock in a mine can be extraordinarily difficult, in part because of the nugget effect. To put it another way, never invest in a gold mine on the basis of just a few assay samples— many samples must be assayed to get an accurate picture of an ore!

Q1.17. In the table, which sample is the outlier, and what is its gold content in ppm?

△ Tool 1.5 The Le Châtelier Principle

Scientists use the term **closed system** to describe some part of the universe that is not exchanging matter with its surroundings. For example, a sugar cube sitting on a table could be considered a "closed system" because the sugar molecules are not wafting away into the air.

However, a drop of water on the same tabletop would not be a closed system. In this case, water molecules leave the drop, evaporating into the air. Note that this evaporation also absorbs a slight amount of heat from the tabletop, cooling it slightly. This is an example of an **open system,** in which both matter and energy move into or out of the system.

Scientists use the term **equilibrium** for systems in which a balance exists between forces or processes such that no net change occurs. For example, consider a cup of hot coffee. It is not at equilibrium for two reasons. Its temperature is not at equilibrium because the coffee is hotter than its surroundings—it is cooling down. Its volume is not at equilibrium because the water is evaporating. Thus, the cup of hot coffee is an open system and it is not at equilibrium because it is experiencing a net change (evaporation of water and loss of heat).

However, suppose we seal the open cup of coffee with plastic wrap. Now it will reach an equilibrium with respect to temperature and volume—eventually. The temperature will reach equilibrium with the coffee cup's surroundings as the coffee cools and the surroundings warm ever so slightly. The volume will reach equilibrium when the rate of evaporation of the coffee equals the rate of condensation dripping from the plastic wrap.

Under equilibrium conditions, the temperature and volume of coffee do not change even though individual water molecules in the coffee are not at rest.

A French chemist named Henry-Louis Le Châtelier developed a valuable concept in the 1800s. It states: "A system at equilibrium will respond to any new change applied to it in such a way as to lessen (minimize) the effect of the change." This is called the **Le Châtelier principle.**

Example: Equilibrium of Water Molecules in the Oceans and Atmosphere. Generally speaking, water neither leaves planet Earth nor is added in significant quantities to our planet from space, so we say that Earth is a closed system for water. Within this closed system, vast volumes of water exist as liquid in the oceans and as water vapor in the atmosphere. The oceans and the atmosphere are in constant contact, so water molecules are free to travel between the two. In one direction, water evaporates from the ocean into the atmosphere, whereas in the other direction, water from the atmosphere falls as rain and snow into the ocean. This exchange of water occurs constantly worldwide.

What is important to us here is that equilibrium exists in the proportion of water molecules that reside in the ocean and those that reside as water vapor in the atmosphere.

Adding heat to the oceans would increase the number of water molecules that evaporate to join the water in the atmosphere. This would disturb the equilibrium. But because the evaporation process absorbs heat energy from the ocean, it also would slightly reduce the ocean temperature, thus minimizing the effect of the change. This is an example of the Le Châtelier principle.

Example: Equilibrium of Sugar Dissolved in and Precipitated from Coffee. Considering a system closer to home, imagine that you stir a large quantity of sugar into a cup of hot coffee. You add so much sugar that not all of it can dissolve, so some accumulates on the bottom of the cup. When you are finished stirring, the sugar crystals at the bottom are in equilibrium with the sugar molecules that are dissolved in the coffee. However, as the coffee cools, it can hold fewer dissolved sugar molecules. This forces the sugar molecules in solution to precipitate, joining the sugar at the bottom of the cup. This precipitation reaction releases heat, thus lessening the effects of the cooling, another illustration of the Le Châtelier principle.

Example: Equilibrium of Solutes in Hot Springs. The situation described in the previous example also exists with dissolved salts in hot springs (Figure 1.11). Where a spring releases water at the ground surface, the hot wa-

Figure 1.11   Hot spring pool with surrounding mineral precipitate, Yellowstone National Park, Wyoming. Note the minerals that have precipitated around the spring as the water cools and evaporates. (*Fritz Polking/Visuals Unlimited*)

ter spreads out. This allows the water to cool, forcing dissolved chemicals to precipitate. An example is silica (the common ingredient in opal, flint, and chert). This precipitation releases heat. It explains the buildup of silica you see around the spring in the photo.

Many precipitation reactions release heat. This makes sense, because you know that in the reverse situation, when you dissolve powdered detergent, salt, or sugar in water, adding heat to the water allows a lot more of the solid to dissolve. This implies that when detergent, salt, and sugar dissolve, they absorb heat.

Of course, other processes can be significant in the chemistry of hot springs. For example, some of the water is evaporating into the air, thus concentrating salts in the water, which also promotes precipitation.

As you will see, the Le Châtelier principle allows us to predict changes that occur when systems at equilibrium are disturbed. Note that the principle doesn't tell us how much heat will be consumed or how much solid will precipitate. It just tells us in what direction changes in the system will move.

Example: Equilibrium of Glaciers. Imagine that Switzerland experiences three unusually warm summers and winters in a row. How can the Le Châtelier principle be used to explain what will happen to Switzerland's famous glaciers? When ice melts, heat is consumed—as you know simply from holding an ice cube in your hand. The melting ice absorbs heat from your hand, leaving your skin quite frigid. Thus, if Switzerland is unusually warm for several years, the local glaciers will respond to lessen this increase in temperature. Glacier ice will melt, a process that consumes heat, and Switzerland's glaciers will shrink, "retreating" up the mountain valleys.

Problem 1.13. Le Châtelier Principle Explains Why Adding Heat Allows Saturated Saltwater to Hold More Salt. Imagine that you have a cold glass of water into which you have dissolved all the salt you can. After stirring and stirring, a thin layer of salt crystals remains on the bottom of the glass. Now imagine that you carefully immerse the glass in a large pan of hot water, thus heating your glass of saltwater. You see that as the water warms up, the salt crystals on the bottom of your glass disappear entirely, dissolving into the warmer water.

Q1.18. How does this event illustrate the Le Châtelier principle?

Problem 1.14. Le Châtelier Principle Explains Whether Diamond or Graphite Forms, Depending on Pressure. Geologists know that the element *carbon* occurs as the soft, slippery mineral *graphite* near Earth's surface, but the same element forms hard *diamonds* deep within Earth. Imagine that a geologist puts a piece of graphite into a tightly sealed cylinder that has a plunger. He then exerts enormous, constant pressure on the graphite. Eventually, he produces a cylinder of carbon that is 90% graphite and 10% diamond. No further change occurs—in other words, the system is in equilibrium under some particular constant pressure.

Q1.19. If at this point the geologist increases the pressure even more, what will happen? How does this illustrate the Le Châtelier principle? (***Hint:*** Diamond is a denser structure of carbon molecules than graphite.)

Unit 1 Paper-and-Pencil Tools

Last (Family) Name _____ First Name _____

Instructor's Name _____ Section _____ Date _____

Q1.1. _____ liters

Q1.2. _____ metric tons

Q1.3. US$ _____

Q1.4. _____ cubic feet

Q1.5. US$ _____

Q1.6. (circle one)   yes   no

Q1.7. porosity unit: _____

Q1.8.
Figure 1.12 Percentage of silver in the gold nuggets.

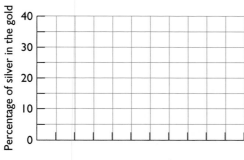

Q1.9. (circle one)   yes   no

Q1.10.
Figure 1.13 Linear plot of exponential population.

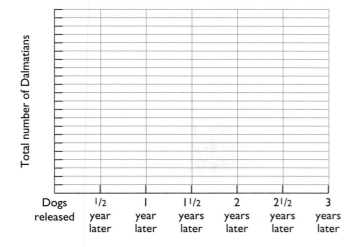

Q1.11.
Figure 1.14   Semilogarithmic plot of exponential population growth of Dalmatian dogs.

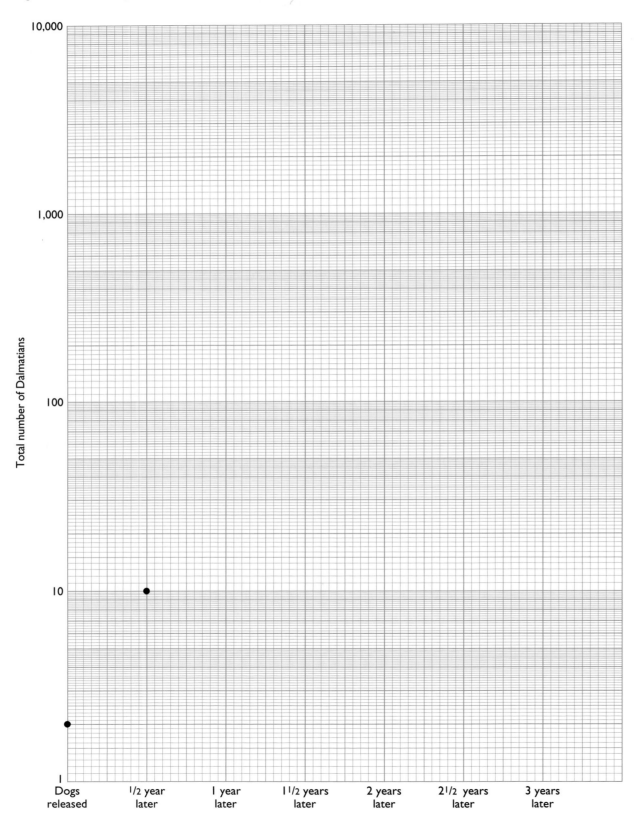

Q1.12. _____

_____

Q1.13.
Figure 1.15 Half-life of uranium-235 shown on linear graph.

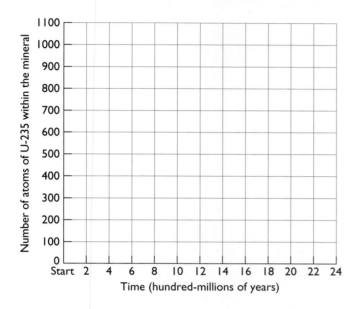

Q1.13. (*cont'd.*)

Figure 1.16 Half-life of uranium-235 shown on semilogarithmic graph.

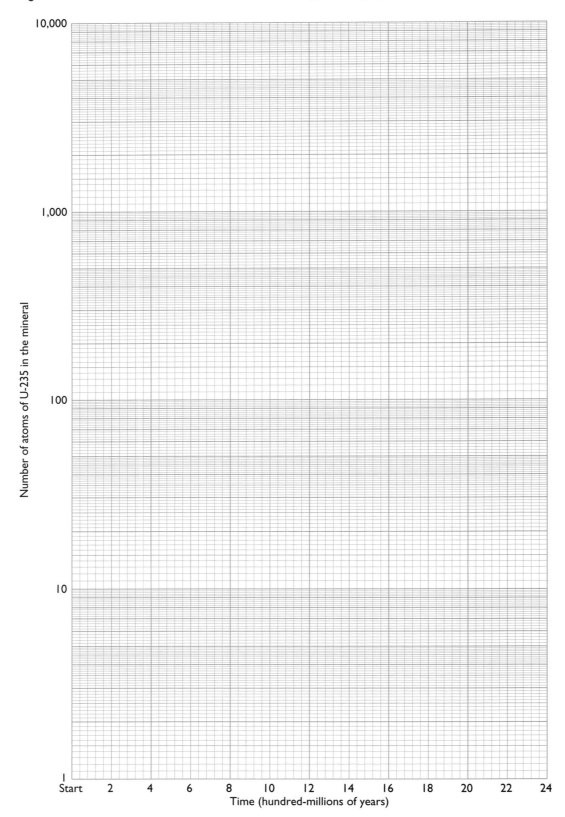

Q1.14.
Figure 1.17 Histogram of grade distribution for the class.

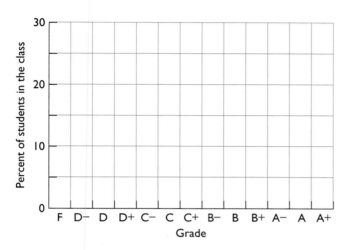

Q1.15. _____ distribution

Q1.16.
Figure 1.18 Histogram showing distribution of gold concentration (ppm) in rock samples.

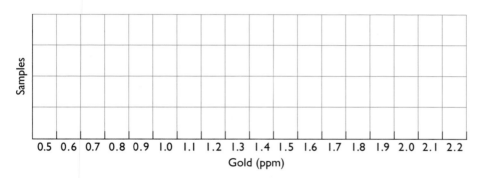

Q1.17. Sample _____ , _____ ppm

Q1.18. _____

_____

_____

Q1.19. _____

_____

_____

This unit presents basic laboratory tools and involves hands-on use of laboratory equipment. You will work with significant figures, interpolation, time, precision, representative sampling, and sieves.

No lab measurement can be perfect. Some error always exists, no matter how good our measuring equipment or how careful we are. When a gem expert weighs a diamond or a geologist weighs a gold nugget, both know that their measurements may be very close to the "truth" but cannot be absolutely accurate.

In these lab exercises, you will determine how *precise* your measurements are. Geologists are a practical lot, and we use a practical approach to deciding how to report the precision of our measurements. You'll be asked to *interpolate* (determine a value between given values) and record measurements in a common-sense manner, indicating the precision of your measurement.

Precision refers to error (variability) in measuring, whereas *accuracy* refers to the error inherent in the measuring equipment. For example, imagine that you are standing on a digital bathroom scale that reports your weight as 136 pounds. If you step off the scale and then back on repeatedly, you may find that 60% of the time the scale reads 136 pounds, whereas 20% of the time it reads 135 pounds and 20% of the time it reports 137 pounds. This reveals that the scale's precision (variability) is ±1 is pound.

When you go to the doctor's office, however, you discover that the doctor's scale, which is more carefully made and maintained, consistently reports your weight at 131 pounds ±$\frac{1}{4}$ pound. This indicates that your bathroom scale reads about 5 pounds high. Thus, your scale has precision of ±1 pound from measurement to measurement, but it has much less accuracy.

Tool 2.1 Significant Figures

Lab 2.1. Measuring Fluid Volume in a Graduated Cylinder. When you measure the volume of water in a graduated cylinder, if you look closely at the top of the water, it will resemble the shape shown in Figure 2.1.

Measure from the lowest point of the *meniscus*, or surface of the liquid. In this case, you can see that the center of the liquid surface is just at the 10-milliliter line. Record your measurement (record only significant numbers) as "10.0 milliliters."

To be more complete, you could add a notation like ±0.1. This tells anyone who reads your data that 10.0 milliliters is the most accurate value you can assign to the volume. It also tells them that you realize that, if there were 0.1 milliliter more or 0.1 milliliter less liquid, you wouldn't be able to see the difference. (***Note:*** The measurement can be written as "10.0 ml," with the last zero denoting the column that has uncertainty within it, or as "1.00×10^1 ml.")

Notice that some containers for liquids automatically make possible more precise measurements due to their design. **Beakers,** or other containers that are short and wide, are relatively insensitive to changes in the volume of a liquid. In other words, adding or removing a little liquid will barely make a visible difference in a short, wide container. A good example is a swimming pool: adding a gallon of water to a swimming pool results in no visible difference in its water level.

In contrast, adding liquid to or removing liquid from a narrow vessel, like a tall cylinder, creates a highly visible rise or fall in liquid level. For this reason, measurements made in tall, narrow **graduated cylinders** are much more precise than measurements done in marked beakers or buckets (Figure 2.2 on the following page). This is why rain gauges are tall and slender, even though the "collector" part may be much wider.

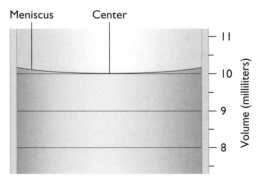

Figure 2.1 Taking a fluid volume measurement when the surface of a liquid is curved: at the center of the meniscus. The volume in this example is 10.0 milliliters.

Figure 2.2 Wide versus narrow containers. In wide containers, adding/removing liquid makes little visible difference in liquid level and is difficult to measure. In narrow containers, adding/removing liquid makes a very visible (measurable) difference.

Materials
- graduated cylinders
- beakers or graduated buckets
- water

Procedure

1. Select a graduated cylinder. Fill it with water. Pour the water into a beaker or bucket. Now refill the graduated cylinder.

2. Compare the depth (height) of the water in each of the containers. Discuss which container enables greater precision in measuring water volume, and the reason why.

3. Add random amounts of water to graduated cylinders and beakers (or graduated buckets) and practice reading the amount, using the meniscus.

Tool 2.2   Interpolation: Volume and Weight

Example: Interpolating the Measured Volume of a Liquid. In most cases, the top of a liquid won't be exactly on a line of a graduated cylinder but somewhere between two lines. This requires you to make a reasonable estimate, or **interpolation,** of the volume indicated (Figure 2.3).

Example: Interpolating the Weight of a Sample (Metric Units). A scale used to weigh material usually presents similar challenges. You may have a scale that uses a dial, like an older, analog (nondigital) bathroom scale. When the needle stops moving, you must interpolate the weight of your sample. Study the examples in Figure 2.4.

Q2.1. Interpolate readings for the two scales shown in Figure 2.5. On the Lab Answer Sheet at the end of this unit, record your readings. Include an estimate of your error in ± form.

Lab 2.2. Measuring and Interpolating the Volume and Weight of Samples. This lab gives you hands-on practice in actual measurement.

Materials
- volume and weight samples **A** though **E**
- graduated cylinder
- mass balance
- water

Procedure
Carefully measure the volume or weight of all samples to make sure you get accurate data.

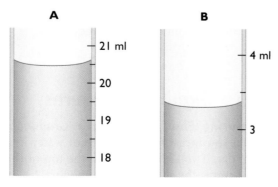

Figure 2.3   Interpolating when reading the level in a graduated cylinder. A. Volume appears to be 20.5 milliliters. **B.** Volume appears to be 3.3 milliliters.

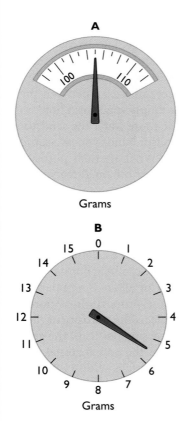

Figure 2.4 **Interpolating when reading a metric system rotary display. A.** Reading is 105 grams ± 1 gram. **B.** Reading is 5.5 grams ± 0.1 gram.

Q2.2. Record your measurements on the Lab Answer Sheet. (**Note:** Always include the units of your measurement.) Tell your instructor your results.

Q2.3. In Table 2.1 on the Lab Answer Sheet, list all of your labmates' measurements along with your own. When you have all the numbers, study them with your labmates. Are any measurements clearly out of line? Can you explain them? Should they be redone? Discuss whether you should discard any measurements, and why.

Once you have agreed on which data are most meaningful and useful, find the average value of the measurements. Record it at the bottom of the table. (**Note:** If you take a class in statistics, you will learn how to respond to this kind of problem in a *quantitative* manner. But for our purposes in this course, a seat-of-the-pants understanding of precision and accuracy in measurements will suffice).

Q2.4. Now that you have your measurements and those of your classmates before you, study them. What uncertainty values do you want to record in your own measurement?

 Tool 2.3   Time and Precision

You now have measured weight (mass) and volume. A third important quantity we will be measuring is time. We need accurate time measurement to understand rate. *Rate* is how much of something happens during a time interval. Some examples that geology students need to determine include:

- the rate at which destructive earthquake waves travel (in feet per second, or kilometers per hour).
- the rate at which contamination in groundwater can travel through the ground toward a water well (in feet per year).
- the rate at which a flash flood moves down a river (in feet per minute or meters per second).

Lab 2.3. Measuring Time and Checking Precision. How accurate is your timing? This lab will help you discover why scientists rely on instruments more than their senses.

Materials
- timekeeping instrument (stopwatch or timer)

Figure 2.5 **Interpolating when reading an English system rotary display.**

Procedure

When your instructor says "Start," start your timekeeping instrument. After a brief period, your instructor will abruptly say "Stop!" Be alert to stop your timer instantly.

Q2.5. Record your time on your Lab Answer Sheet, being sure to note the units and to indicate an estimate of error.

Q2.6. Pool your datum with the data of your labmates. Record the values in Table 2.2 on your Lab Answer Sheet. Are all the measurements meaningful, or did anyone miss the mark significantly? Which data should be retained, and which should be dropped? When these questions are answered to everyone's satisfaction, find the average value of the "good" measurements and record it in the table.

Tool 2.4 Representative Sampling: Using Point Counts

Scientists who work in the field (in other words, outside the lab) must decide which specific samples to collect. For example, if you are a geologist in the field, you might need to collect a rock sample from an outcrop to take back to the lab for study. Obviously, you must choose a sample that is representative of the whole outcrop, not one that is unlike most of the outcrop. So what sample should you collect? How big? And from where in the outcrop?

If the outcrop is uniform in every way you can see—let us say that it is all one solid, uniformly colored, uniformly textured granite—then a single 1- to 3-kilogram sample of the rock might be representative. But if the outcrop is visibly varied in color or texture or in the severity with which it has weathered, you might have to collect several samples to roughly represent the rock at the whole outcrop.

Another consideration is scale. If you stand 50 feet back from an outcrop, it may all look the same. But if you study it up close, say from 3 feet away, you may see significant variation in the rock on a small scale. Thus, how many samples of the outcrop you need—one, a few, or many—may depend on the scale of the features you wish to measure.

Lab 2.4. Representative Sampling and Point Counts. In the lab, the need for representative samples may be critical. For example, someone might want to drill a well into sandstone to obtain a water supply for a new home. Suppose you are the geologist assigned to determine how much water the sandstone can hold. To do so, you must know the *porosity* of the sandstone: the volume of the tiny empty spaces that exist between the sand grains in the rock. These empty spaces can hold water. The greater the porosity, the more water the sandstone can hold.

Your supervisor hands you two samples of the sandstone and asks, "What are their porosities?" (Let's hope that your supervisor obtained truly representative samples of the sandstone!)

One way to determine the porosity of the samples is to look through a microscope at the sandstone and count its tiny pores. This is easily done by superimposing a grid on the sample (Figure 2.6). You then count the pores that happen to lie at the crossing points of the grid. Knowing the total number of all crossing points on the grid, you can determine the proportion of pores in the sample. This number reflects the porosity of the rock.

Using the crossing points of the grid lines on Figure 2.6, check whether the rock material immediately below each crossing point falls within a pore or falls on solid rock. Count how many crosses are in pores and how many are in rock. If you think the crossing point is near a grain boundary, use your own judgment about whether it's closer to the grain or the pore.

Q2.7. As you proceed, record your work in Table 2.3 on your Lab Answer Sheet. Using your counts, calculate the area of the samples that corresponds to pores. Express your answer as a percentage: pore counts divided by total counts.

Q2.8. How could you change the grid to make your answer more accurate?

In many cases, the percentage of the area of a rock sample that is composed of pores is equal to the volume of the whole rock that is three-dimensional pores. Thus, the point-count method gives a reliable indication of the sandstone's porosity and how much water it could hold. However, your instructor may indicate circumstances where this assumption does not work. If so, note here:

_____

_____

Tool 2.5 Using Sieves

Let's say you are babysitting a child, playing with her in a sandbox. You lose a small earring in the sandbox. What is the quickest way to find it? Use a sieve! A **sieve**

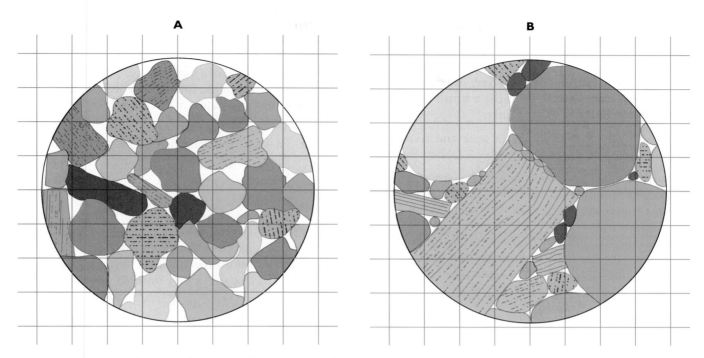

A

B

Figure 2.6   Porosity of two sandstone samples—coarse grid. A. Quite porous. **B.** Not very porous.
The porosity is revealed by superimposing a grid of squares over photomicrographs of two
sandstone samples.

is simply a screen, like a window screen. Its holes allow particles that are smaller than the holes (the sand grains) to fall through but larger particles (like the earring) to be retained on the screen, thus separating particles of two different size classes.

Geologists often need to sort loose materials by size —for example, sands and gravels. A real-world example might be in a gold field, where geologists may know from experience that gold particles in a streambed are all of a similar size. Thus, to recover the gold most efficiently, a geologist will sort all the loose material in the streambed by size.

Generally, such sorting is done with a set of several sieves. Each sieve has a particular spacing between its wires. The spacing usually ranges from about 3 to 5 millimeters (the thickness of two or three quarters) to less than 0.5 millimeter (thinner than a pin). The results are then displayed in a histogram (review Unit 1). **Lab 2.5. Sieving a Sample, Recording Results, and Constructing a Histogram.** Your instructor will give you a sample of gravels and sands and ask you to determine the distribution of particle sizes. How can you accomplish this? You could hand-sort the sample, grain by grain, but it would take a very long time. Instead, if you use sieves, a weight scale, and a little arithmetic, you will obtain the same result quickly and more accurately.

Materials

- sets of sieves
- rulers for measuring the size of the holes in the sieves
- samples of mixed sands and gravels
- mass balances

Procedure

1. Weigh your entire sample.

Q2.9. Record the total sample weight in the bottom cell of Table 2.4 on the Lab Answer Sheet.

2. In the second column, write the coarsest particle size that can pass through each sieve. (Use inches or millimeters, and indicate which one you used.)

CAUTION   POTENTIAL EQUIPMENT DAMAGE: Please handle sieves with care! The screens are fragile.

3. Sift the material through the sieves. Start with the coarsest sieve and progress through the next-finer and next-finer sieves. Depending on your sieves, you may have to do this one size fraction at a time. Or you may be able to stack your sieves and do several size fractions at once.

Generally, a gentle shaking motion will help the particles sort and fall. Be patient! Patience is a virtue in sieve work, as in all lab tasks.

4. When your sample is fully sorted, carefully remove the material from each sieve, using a brush or spatula. Again, please treat the sieves gently.

5. Weigh each size fraction on a scale and record the weights in the fourth column of Table 2.4. Show the unit.

6. Sum the weight values of all the size fractions. Record your result in the cell labeled "Arithmetic total of weights." Show the unit.

Q2.10. Find the difference between total sample weight and the arithmetic total of weights (subtract). Record your results.

Q2.11. Divide this difference into the total sample weight and record. This is your error estimate.

7. Check with your lab instructor to see if your work at this point is satisfactory.

8. Convert your weight data into percentages. To do so, divide each value by the total sample weight. Record the results in the last column of Table 2.4.

Q2.12. Now you can construct a histogram of your results using Figure 2.7 on your Lab Answer Sheet.

A sieve allows you to measure particle size, but particle shape is a factor, too. Obviously, nicely rounded particles pass right through. But is there a particle shape that you might not expect to pass through a particular sieve, yet it did, or could have done so?

Q2.13. Describe the shape.

Unit 2   Laboratory Tools

Last (Family) Name _____ First Name _____

Instructor's Name_____ Section _____ Date _____

Q2.1.

A. _____ ± _____ounces

B. _____ ± _____ounces

Q2.2.

Sample A: _____ _____ (number and unit)

Sample B: _____ _____ (number and unit)

Sample C: _____ _____ (number and unit)

Sample D: _____ _____ (number and unit)

Sample E: _____ _____ (number and unit)

Q2.3.

Table 2.1

Measured values (weight and volume) for Samples A-E

Student	Sample A	Sample B	Sample C	Sample D	Sample E
Total:					
Average:					

Q2.4. _____

_____

Q2.5. Time measurement (be sure to include unit): _____

Estimated error:_____

Q2.6.

Table 2.2

Measured values for time

Student	Time	Estimated Error
Total:		
Average:		

Q2.7.

Table 2.3

Point counts: sandstone porosity, coarser grid

	Sample A	Sample B
Pore counts		
Rock counts		
Total counts		
Area of pores expressed as a percentage		

Q2.8. _____

_____

Q2.9.

Table 2.4

Particle size distribution

Particle size fraction	Particle diameter (indicate millimeters or inches)		Weight of each size fraction (indicate grams or ounces):	Percentage of total weight:
	Greater than	**Smaller than**		
1. Coarsest		—		
2. Next-finer				
3. Next-finer				
4. Next-finer				
5. Next-finer	—			
Arithmetic total of weights:				100
Total sample weight:				

Q2.10. The difference is _____. Show the unit.

Q2.11. _____ %.

Q2.12.

Figure 2.7 Histogram of particle size ranges. Distribution of five different particle size ranges determined by sieving a sand sample.

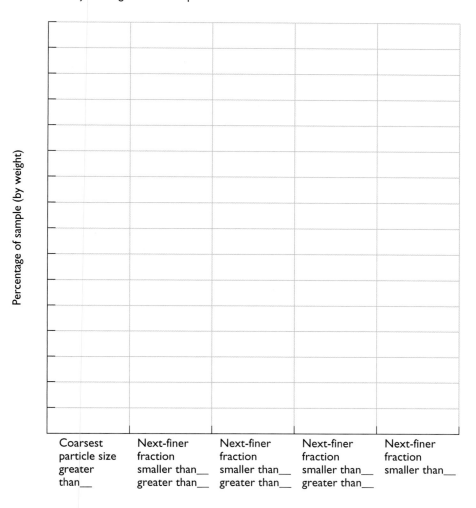

Coarsest particle size greater than__

Next-finer fraction smaller than__ greater than__

Next-finer fraction smaller than__ greater than__

Next-finer fraction smaller than__ greater than__

Next-finer fraction smaller than__

Percentage of sample (by weight)

Q2.13. _____

Geologists study the solid materials of Earth—minerals and rocks. Some minerals have economic value, such as rubies, emeralds, and graphite. So do some rocks, like granite, limestone, and coal. But all rocks and minerals, no matter how valued or humble, are of interest to geologists who study the genesis and evolution of our planet. Rocks and minerals that compose them are the topic of Units 3–6.

Minerals and Gems   Unit 3

All minerals are solid **crystals,** in which the atoms are arranged in a highly ordered structure. Most minerals have a definite chemical composition. Earth's minerals include rare gems like diamond, ruby, emerald, and sapphire. More common minerals include graphite (the black substance in pencils), halite (table salt), garnet, and quartz.

This unit introduces you to the laws of chemistry and mineralogy that govern how minerals form and what properties they have—their solubility, hardness, shape, color, strength, taste, and so on. We will guide you through simple experiments in crystal growth, mineral solubility, mineral hardness, and more. Then you'll apply this knowledge to recognizing small samples of common minerals.

Minerals are the building blocks of rocks. Thus, this unit on minerals is followed by three units on rocks—igneous rocks, sedimentary rocks, and metamorphic rocks. All are composed of the minerals you study in this unit. Because rocks are composed of minerals, it is important that you learn how to recognize each mineral and its name. You'll also see how this knowledge has application throughout the real world.

Lab 3.1   Solubility of Minerals

In this pair of experiments (Part A, Part B), you will dissolve some common minerals, first in water and then in a weak acid.

Part A   Solubility of Halite, Sylvite, and Gypsum in Water

Materials

- three small beakers or cups
- water
- measuring spoon
- halite (ordinary table salt, chemically called *sodium chloride*)
- sylvite (a dietetic salt substitute, chemically called *potassium chloride*)
- gypsum (ground or powdered)
- watch or clock that shows seconds

Why do some minerals dissolve in water? The answer lies in the chemical bonds that link atoms within a crystal. For minerals that dissolve, the bonds within the crystal are weak enough to be overcome by the opportunity for stronger bonding with water molecules. Water (H_2O) is an unusual solvent. The hydrogen atoms of the water molecule have a net positive charge, while the oxygen atom has a net negative charge. This makes water a "polarized" molecule, and the negative and positive "ends" of a water molecule can attract the atoms that make up a mineral. In terms of chemical symbols, a solubility reaction for table salt looks like this:

$$NaCl + H_2O \rightarrow Na^+ + Cl^- + H_2O$$

solid   water   dissolved   dissolved   water
salt   sodium   chloride
                ion   ion

As you will see, the amount of material that dissolves into water during a finite time also can depend on the dissolution rate of different minerals.

Procedure

A1. Make sure the three beakers are clean and dry.

A2. Carefully measure 10 spoonfuls of water into each beaker. (The water must be at the same temperature in all three beakers; do not use warm water for one and cool water for another.) Dry the spoon completely.

A3. Two of the mineral samples are sold in grocery stores. One is the mineral **halite**—regular table salt, or sodium chloride. The other is **sylvite,** a "salt substitute" used in foods, or potassium chloride. With clean hands, shake out a few grains of each type of salt and note the difference in their taste.

A4. Measure one level spoonful of the regular salt (halite). Stir it into the first beaker of water, stirring continuously until it all has dissolved. Time how many seconds it takes.

Q3.1. In Table 3.4 on the Lab Answer Sheet, answer the question in the second column and record your time.

A5. Repeat the experiment with a spoonful of sylvite (the salt substitute) in the second beaker. Record your observations in the table. (**Hint:** Pure sylvite will dissolve entirely, and rapidly, but a tiny amount of a white, insoluble mineral may be included in the salt substitute. It's added to keep the salt flowing freely, even in humid conditions, and perhaps to partially cover the flavor of the sylvite. Ignore this tiny residue, which will not dissolve.)

A6. The third sample is pulverized **gypsum,** a common mineral composed of calcium, sulfate, and water. This is the same material you would have if you mixed powdered plaster and water and let it set (harden). Repeat the experiment with a spoonful of the ground gypsum in beaker 3. (You can stop looking after a couple of minutes have passed.) Record your results in the table.

Applying What You Just Experienced

Think of how water from rain and snow runs endlessly over rocks and through soil, eventually draining into the ocean. It's the reason that seawater contains a lot of material that has been dissolved from soluble minerals in rocks and soil. In fact, the chemical composition of seawater is dominated by six substances:

chlorine (Cl)
sulfate (SO_4)
sodium (Na)
potassium (K)
calcium (Ca)
magnesium (Mg)

Look at the following chemical formulas for the mineral samples:

halite (NaCl)
sylvite (KCl)
gypsum ($CaSO_4 \cdot 2H_2O$)

Q3.2. How many of the six dominant substances in seawater are present in the mineral samples?

Q3.3. Why can we describe the ocean as containing a lot of dissolved gypsum, when you observed that little gypsum dissolved in your container? (**Hint:** There might be several answers, but consider rate.)

Not so long ago, the U.S. Army Corps of Engineers built a dam in Utah. A lake formed behind it, and the lake established a shoreline. But soon the water in the lake drained away! Geologists found that the dam had been built into rock that actually was a thick layer of gypsum. Gypsum is a little slow to dissolve, but it certainly dissolves over time! The gypsum around the dam dissolved, leaving a pathway through which the lake water simply drained away. The dry-land dam became a costly monument to geologic ignorance.

Q3.4. What type of scientist should be consulted before engineers start to build a dam?

Part B   Solubility of Calcite, Dolomite, and Quartz in Weak Acid

Only a few minerals, like halite, are highly soluble in pure water. Most minerals do not noticeably dissolve in water, even after weeks or months. But in a weak acid, certain minerals dissolve and release carbon dioxide gas as they do so. This happens because the acid breaks the bonds that hold the mineral's atoms together within the mineral's crystal structure. Acid contains hydrogen ions (H^+), which are particularly good at breaking certain chemical bonds in minerals. Carbonate minerals, in particular, dissolve in weak acids, releasing carbon dioxide gas.

For example, a gorgeous rose-pink carbonate mineral called *rhodochrosite* ($MnCO_3$) will dissolve in a weak acid like this:

$$MnCO_3 \;+\; H^+ \;\rightarrow\; Mn^{+2} \;+\; HCO_3^-$$

solid   hydrogen ion in water   manganese ion in water   carbonate ion in water

$$\rightarrow\; Mn^{+2} \;+\; CO_2 \;+\; OH^-$$

manganese ion in water   carbon dioxide gas bubbles   hydroxyl ion in water

Materials

- weak acid in a dropper bottle
- samples of calcite, dolomite, and quartz

Procedure

SAFETY CHECK ➤ The acid used in this experiment is diluted with water to make it weak, but be careful. Keep the acid away from your face. If some acid splashes on your clothes, it may bleach the color out of the fabric. Wash your hands if you spill acid on them and immediately flush your eyes with water if you get *any* acid in them.

B1. You have been given samples of three common minerals, **calcite, dolomite,** and **quartz.** One by one, place a drop of acid on each sample.

Q3.5. Record what you see and hear in Table 3.5 on your Lab Answer Sheet. (Are there bubbles in the liquid? A lot? A few? Can you hear fizzing?)

B2. Look up the formulas (chemical compositions) of the three minerals in your textbook.

Q3.6. Write their formulas on your Lab Answer Sheet.

Q3.7. Why might the dolomite and calcite behave in roughly similar ways in the presence of acid?

Q3.8. Why did your quartz sample not produce any carbon dioxide gas bubbles?

Applying What You Just Experienced

Silicate minerals are not very soluble, even in weak acids, due to the powerful bonds among the silicon and oxygen atoms in silicate mineral crystals. However, carbonate minerals are another story. All carbonate minerals have relatively weak bonds, so they dissolve in even the weakest acids. This dissolution is commonplace because ordinary rainwater generally is slightly acidic.

The natural source of the acidity in rainwater is carbon dioxide gas. Carbon dioxide is a natural part of Earth's atmosphere, and it dissolves into raindrops to form weak carbonic acid. The chemical equation that describes this process looks like this:

$$H_2O \;+\; CO_2 \;\rightarrow H_2CO_3 \rightarrow\; H^+ \;+\; HCO_3^-$$

water   carbon dioxide gas   carbonic acid   hydrogen ion   carbonate ion

Another source of acid in raindrops is industrial pollution. The burning of coal and oil releases airborne sulfur and nitrogen compounds. These compounds combine with water in raindrops to form dilute sulfuric acid and nitric acid. Thus, *acid rain* is a significant problem wherever industrial pollution occurs. In China, where a great deal of "dirty" coal is burned (coal containing more sulfur impurities than allowed in U.S. and Canadian power plants), some raindrops are quite acidic.

Wherever acid rain falls, carbonate minerals slowly dissolve. For example, the reaction for calcite looks like this:

$$CaCO_3 \;+\; H^+ \;\rightarrow\; Ca^{+2} \;+\; HCO_3^-$$

calcium carbonate (calcite)   hydrogen ion in rain   calcium ion in solution   carbonate ion in solution

The reaction for dolomite looks like this:

$$CaMg(CO_3)_2 \;+\; 2H^+ \;\rightarrow Ca^{+2} \;+\; Mg^{+2} \;+\; 2HCO_3^-$$

dolomite   hydrogen ion in rain   calcium ion in solution   magnesium ion in solution   carbonate ion in solution

Marble is made of a carbonate mineral, namely, calcite. Marble is popular for its beauty in statuary and tombstones. However, the effect of acid rain on marble is quite noticeable, even within a human lifetime. A hundred-year-old marble tombstone in a humid climate displays names and dates that look shallow and eroded.

Q3.9. Considering what you have just learned, if you wanted your own tombstone to remain legible in the year 3000, would you have it made from a rock made of a carbonate mineral? Why or why not?

Lab 3.2 Hardness of Minerals

The arrangement of atoms within a crystal is controlled by the chemical bonds that hold the atoms in place. Some bonds are stronger than others. This fact, along with crystalline structure, determines the **hardness** of each mineral. Hardness can be quite helpful in identifying a mineral. Minor chemical impurities in a mineral can change its color and thus confuse you as to its identity, but hardness varies only slightly with minor impurities.

Materials
- mineral samples labeled **A, B, C, D, E, F**
- penny
- glass plate
- steel nail or knife
- your fingernails
- streak plate

Procedure

1. At first glance, four of the mineral samples (**A, B, C, D**) may look similar because they are clear or translucent. But try rubbing a corner of one mineral against the face of another. Note how they differ in hardness.

Q3.10. Try rubbing each of them against the other, and you'll discover that one of the four will scratch the other three but not be scratched by them. This is the hardest mineral. Enter its letter (**A, B, C,** or **D**) in the appropriate cell of Table 3.6 on your Lab Answer Sheet.

(*Note:* When you try to scratch one mineral with another, tiny particles of broken material may leave a streak on the harder mineral. But this is just powder, not a scratch. Try rubbing off the mark with your finger to be sure you see a true scratch (a tiny trench), not just a powder mark left by a softer mineral. Also, always try to use a corner or point on one specimen to scratch another. A flat face pushed against a flat face will not produce the desired scratch.)

2. Continue experimenting to create a rank ordering of the hardnesses of the four samples. Complete Table 3.6 on your Lab Answer Sheet.

Geologists use the **Mohs hardness scale** to describe the hardness of minerals. The scale is named for Friedrich Mohs, who developed it early in the 1800s (see your textbook). This scale uses actual minerals as hardness standards, running from 1 (the mineral talc, which is so soft that it is used as baby powder) to 10 (diamond, the hardest natural substance on Earth). Table 3.1 and Figure 3.1 show this scale.

The hardness of the same mineral can vary slightly because *chemical substitutions* occur within mineral crystals. Also, hardness may vary somewhat with direc-

Table 3.1

Mohs scale of hardness for some minerals

Hardness value	Mineral
10	Diamond (hardest mineral on Earth)
9	Corundum (including the varieties known as ruby and sapphire)
8	Topaz and beryl (including the variety called emerald)
7	Quartz
6	Orthoclase (potassium) feldspar
5	Apatite
4	Fluorite
3	Calcite
2	Gypsum
1	Talc (used to make talcum powder because of its softness)

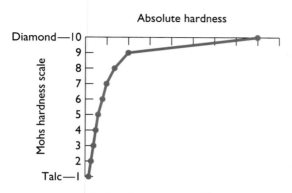

Figure 3.1   Mohs hardness vs. absolute hardness.

tion in a single specimen—it may be easier to scratch the specimen in one direction than another.

Many minerals have a hardness value that lies between the whole numbers listed in the table. For example, halite (rock salt) has a hardness of about 2.5 on Mohs scale, and garnets range in hardness from 7 to 7.5.

Hardness is a useful property that helps us identify minerals. The procedure is simple: a scratch test, like the one you performed at the start of this exercise. To conduct scratch tests on unknown minerals, you must use materials of known hardness, like the common household materials listed in Table 3.2.

3. Using the common materials supplied, plus one you supply yourself (your fingernails), scratch-test mineral samples **E** and **F**. Also, determine the Mohs hardness values for the first set of samples you used (**A, B, C, D**).

Q3.11. Use your results to complete Table 3.7 on your Lab Answer Sheet.

Applying What You Just Experienced

A mineral's hardness depends on its **crystal structure** as well as the elements it comprises. A classic illustration of the importance of structure in determining a mineral's hardness is a comparison of diamond and graphite. Both are made of identical carbon atoms. Yet diamond is the hardest mineral on Earth (making it useful in cutting tools and as durable jewelry), whereas graphite is one of the softest minerals (making it usable as pencil "lead" and as a lock lubricant).

Geologists say that diamond and graphite are *polymorphs* (Greek: "many forms") because they have the same composition but different crystal structures. It is their different crystal structures that give them their different physical properties. Because crystal structure is determined by chemical bonds, you can see that diamond must have very strong bonds in all directions, whereas graphite has some very weak bonds.

Hardness is one of the most important properties of minerals used by geologists for identification purposes. This is especially true for the common minerals you will be handling in lab.

Thus, the Mohs hardness scale is very useful for mineral identification. But note that it is quite an unusual scale, scientifically speaking. Diamond (Mohs hardness 10) is *not* ten times harder than talc (hardness 1). Fluorite (hardness 4) is *not* half as hard as topaz (hardness 8). Thus, the Mohs scale is not a simple ratio scale, so we cannot use it for calculations. Graphically, this idea is presented in Figure 3.1. If the Mohs hardness scale were linear, the points drawn in the figure would be linked into a single straight line in Figure 3.1.

Q3.12. Judging from Figure 3.1, about how many times harder (absolutely) is diamond compared to corundum? Diamond compared to talc?

Table 3.2	
Mohs hardness values of common materials	
Material	**Mohs hardness**
Streak plate (unglazed ceramic or porcelain tile)	about 7
Steel file (steel that has been "hardened")	6.5
Ordinary glass (windows, bottles)	5.5
Knife blade (hardness of steel varies)	5 to 6
Common "wire" nail	about 4
U.S. penny (copper coin)	about 3
Your fingernail	just over 2

Clearly, the Mohs scale is not a linear one. But the higher the value on the Mohs scale, the harder the mineral. Although geology is a fully modern science, it still uses some old-fashioned (even quaint) words and scales. But the nonlinear nature of the Mohs system does not hinder doing quality work.

Lab 3.3 The Way Minerals Break (Cleavage)

The atomic structure of a mineral—the way atoms are arranged in its crystal lattice—determines the mineral's shape. This is true both for the mineral's *crystal form* (the shape into which it grows) and its **cleavage form** (the shape into which it breaks). Note that these two shapes, crystal form and cleavage form, can be quite different. A crystal may grow into one particular shape, but broken pieces of the mineral usually have a different shape due to weaknesses in bond strength in a particular direction on a plane. (See your textbook's discussion of cleavage.)

Crystals often break so that the pieces have one or more smooth, flat sides. These sides are so uniform that each one forms a surface smooth as a tabletop, and light will reflect off it at a certain angle. When geologists see this reflected light, they say that the smooth cleavage surface is "winking" at them.

The following are the most common and easily studied cleavage shapes:

- *Basal:* cleavage in one direction, forming thin sheets, like a stack of papers
- *Prismatic:* cleavage in two directions, typically not at right angles, and no third side
- *Cubic:* 90° angles between cleavage surfaces and cleavage in three directions, like a cube or a shoebox
- *Rhombohedral:* similar to cubic cleavage, but angles are not at 90°, so the shape is that of a "tipped" or "squashed-over" box
- *Octahedral:* cleavage in four directions, producing an eight-sided shape, like two four-sided pyramids joined bottom to bottom.

Examples of cleavage types are shown in Figure 3.2. Some minerals have no preferred shape when broken, so geologists say they have fracture but no cleavage. If they

Basal

Cleavage in one direction

Prismatic

Cleavage in two directions   Cleavage in two directions
at right angles   not at right angles

Cubic   **Rhombohedral**

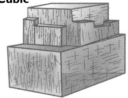 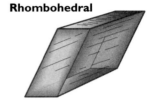

Cleavage in three directions   Cleavage in three directions
at 90°   not at 90°

Octahedral

Cleavage in four directions

Figure 3.2 Cleavage of common minerals. (*Drawings by L. Davis*)

break like glass, forming concentric rings, like a seashell, we call the fracture **conchoidal**.

Materials

- mineral samples labeled **G, H, I, J, K, L**
- hammer
- safety goggles; small plastic bags

Procedure

SAFETY CHECK ▶ **Be careful and gentle with the hammer! Strength is not needed—precision is.**

Wear safety glasses or goggles or put the mineral sample in a plastic bag, to prevent fragments from flying, if your instructor so directs.

1. Gently break sample **G** with the hammer. Study the resulting small pieces. Note that all have the same basic shape. Select three pieces, choosing them to be as different from one another as possible.

Q3.13. Sketch the three pieces on your Lab Answer Sheet.

2. The other samples, marked **H** through **L**, already have been broken for you, so do not break them further. Study the samples with care. (Your instructor may also provide tiny pieces of these minerals. In real rocks, most mineral pieces are small, so study the small pieces diligently, looking for the characteristic cleavage shapes.)

Q3.14. Complete Table 3.8 on your Lab Answer Sheet.

Applying What You Just Experienced
Cleavage is a physical property of minerals. Like other physical properties, it depends on the way atoms are arranged within the crystal. Consequently, we can deduce something about the invisibly small atomic structure of a mineral by the way it breaks!

Q3.15. Knowing this, how would you describe the atomic structure of mineral **G**?

Lab 3.4 Mineral Color versus Streak Color

Materials
- four samples of hematite labeled **M, N, O, P** (although all four are the same mineral, the samples all look different)
- streak plate (white, unglazed porcelain tile)

Procedure

1. The samples are of the iron oxide mineral *hematite*. (Its typically red color was likened to blood, so its name comes from an ancient word for blood, as in *hemo*globin). As you can see, hematite can vary across a range from a gray, metallic luster to a red, earthy luster.

Q3.16. In Table 3.9 on your Lab Answer Sheet, describe the luster and color of the samples. (See your textbook's discussion of luster.)

2. Scrape a corner of sample **M** across the white streak plate. Press hard, making a single strong, clear mark across the tile. Now gently rub off any loose fragments of mineral on the tile. In the last column of Table 3.9, describe the color of the streak as accurately as you can. Use several words or phrases if necessary.

3. Do the same for the other samples and complete the table.

Applying What You Just Experienced
Geologists and rock hounds know that most minerals leave a white or pale streak on a streak plate. A strongly colored mark is unusual, so it helps to identify the few minerals that have a richly colored streak, like hematite.

Note that all your hematite samples have red in their streaks. (If you're not convinced, look closer at the streaks and note the red within the brownish red and grayish red marks. If you did not use "red" or "reddish" in any of your hematite streak descriptions, now is the time to amend your answers in the table.) It takes practice to see minor differences in color and know the significant colors of streak.

Q3.17. If you encounter an unknown mineral later in this lab that gives a reddish streak, what mineral name should you consider assigning to it?

Lab 3.5 Density of Minerals

Density is weight per unit volume. The density of minerals varies over a wide range because some minerals are made of heavy elements (like lead) and because some crystal structures pack together the atoms more tightly than other structures. In either case, it is the atoms within a mineral and their arrangement in the crystal's structure that determine the mineral's characteristic physical properties.

In this lab, you will determine the density of some common minerals. Most minerals have a density near 2.7 grams per cubic centimeter (g/cm^3), so you unconsciously consider this value "normal" when you heft a mineral or rock sample. But, as you will see, some minerals have distinctly different densities.

Materials

- three minerals labeled **Q, R, S**
- graduated cylinder
- water
- mass balance

Procedure

1. Weigh each mineral sample on a mass balance. Use grams (or convert any other weight units used into grams, using the method of unit conversion in Unit 1).

Q3.18. Record the measurements (and an estimate of error) in Table 3.10 on your Lab Answer Sheet.

2. Half-fill a graduated cylinder with water (no need to be exact). Record the volume of water you have (use milliliters): _____. (If the volume lies between the graduations on the cylinder, you will have to interpolate the value as shown in Unit 2.)

3. Tipping the cylinder to one side, *gently* slide mineral **Q** into the cylinder so that it moves to the bottom of the cylinder. (If any water splashes out, start over with step 2.)

4. Placing the graduated cylinder on a flat table, tap out any large air bubbles that may be under or around the mineral. Record the volume that now is indicated by the surface of the water: _____.

5. Subtract your first reading from your second to find the volume of water displaced—as you did in Unit 2. This gives the volume of mineral **Q**. Record that value in Table 3.10, showing your estimate of error in the measurement.

6. Repeat this procedure for minerals **R** and **S**. Add this information to Table 3.10, again with an estimate of your measurement error.

7. Density is the weight of a sample divided by the sample's volume. Written as a formula:

$$\text{density} = \frac{\text{weight}}{\text{volume}}$$

For units, geologists think of density as *grams per cubic centimeter,* which is the same as *grams per milliliter* because 1 milliliter = 1 cubic centimeter. Using the techniques you learned in Unit 1, calculate the densities of the three mineral samples and enter them in Table 3.10.

8. Careful measurements by mineralogists indicate that the densities of the minerals in question are:

Mineral **Q**: 2.7 grams per milliliter

Mineral **R**: 5.0 grams per milliliter

Mineral **S**: 7.4 to 7.6 grams per milliliter

(***Note:*** Because water has a weight of 1 gram per milliliter, the mineral density values are equivalent to what mineralogists call *specific gravity,* which is the ratio of a mineral's density compared to the standard density of water, or 1.0. Your textbook may discuss density in terms of specific gravity.)

Q3.19. Your table of values doubtless differs from the mineralogists' answers. What three or four factors might contribute to the difference between your values and the standard values of these minerals' densities? (***Hint:*** For one of the less obvious reasons, notice that mineral **S** is recognized by mineralogists as having a range of densities rather than a single value.)

Applying What You Just Experienced

Density, like the other properties of minerals, depends on both composition and atomic structure. Mineral sample **S**, for example, is composed of lead and sulfur atoms, and is called *galena.* Its greater density reflects that one of its constituents, the metal lead, is a heavy element. Mineral **R** (pyrite, or fool's gold) is less dense because it is made of iron and sulfur atoms. Least dense of the group is mineral **Q,** quartz, which is made of lighter silicon and oxygen atoms not closely packed.

Q3.20. Judging by the "heft" of your galena sample, is it the most dense mineral you have handled during any part of this lab?

As you become more familiar with common minerals, you will discover that density ("heft") is useful in distinguishing among some minerals.

(***Safety Note:*** Because galena contains lead, you should wash your hands with soap and water before you leave lab today.)

Lab 3.6 Optical Properties of Minerals

The highly ordered structure of minerals affects the passage of light through them. The structure even affects light reflected from the surface of a crystal. Again, it is the atom-by-atom arrangement in the crystal structure that determines the properties of minerals, including how they transmit and reflect light.

This experiment has four parts—

Part A   Refraction
Part B   Polarization and Rotation
Part C   Double Refraction
Part D   Directional Color Change

Part A   Refraction

Materials

- ruler
- protractor
- cubic piece of clear mineral labeled **T**
- low-power laser light

Procedure

SAFETY CHECK ➤ Don't look directly into the laser light, and do not point it at your labmates. Even this low-power unit can damage human eyes due to the heating effect of the laser on tissues. Recall that higher-powered lasers are used to burn holes in materials!

A1. Place the cubic crystal (**T**) on a piece of paper. Point the laser at a 45° angle along the edge of the crystal. Do this in a manner that allows the laser light to travel directly to the paper, where it will make a red dot (red dot 1 in Figure 3.3), not passing through the crystal at all.

A2. Now move the laser slightly, while keeping it at the same 45° angle, so that the light *does* pass through the crystal. Note that the light is bent, or *refracted*, by the crystal, and it reaches the paper at a different angle (red dot 2 in the diagram).

A3. Using a ruler marked in millimeters, measure the lengths *a, b,* and *c* of the crystal, as shown in Figure 3.3.

Q3.21. Make a scale drawing of your crystal, using all your measurements, in Figure 3.12 on your Lab Answer Sheet.

Q3.22. Using the protractor, measure the difference between the two angles, α (alpha) and β (beta). How many degrees has the laser light been bent, or *refracted*, by its passage through the crystal?

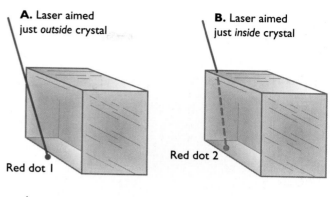

A. Laser aimed just *outside* crystal

B. Laser aimed just *inside* crystal

Red dot 1

Red dot 2

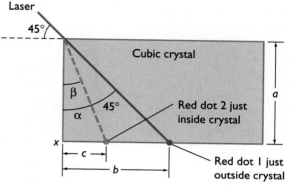

Laser

45°

Cubic crystal

β

45°

α

Red dot 2 just inside crystal

a

x

c

b

Red dot 1 just outside crystal

a = height of crystal
b = horizontal distance from red dot 1 to point x, which is directly below where laser beam passes beside upper edge of crystal
c = horizontal distance from red dot 2 to point x, which is directly below where laser beam enters crystal

Figure 3.3   Laser refraction in a halite crystal.

Part B   Polarization and Rotation

Materials

- minerals labeled **T, U, V, W, X**
- small vial of water
- two pieces of polarizing film

Procedure

B1. Examine the two pieces of polarizing film. They are just like the material in polarized sunglasses. (A polarizer is like a coin slot. You have to insert the coin at the proper angle for it to go into the slot. If the coin is at any other angle, it can't go in. With polarized film, light waves that are lined up in the proper direction pass through the film. Light waves that are lined up at different angles are blocked.)

B2. Look through either of the films and you will see that the world around you looks darker. Now place the

A

B

Figure 3.4 **A. Crossed polarizers.** Each piece of film acts as a coin slot to the passage of light, letting light of only a certain polarization pass through (just as a coin must be oriented properly to go into a coin slot). Here, two pieces of polarized film are held so that one is turned 90° with respect to the other, blocking nearly all light from passing through (like putting two coin slots together at 90° so that a coin cannot pass through them). **B. Two transparent minerals placed between two polarized films.** The films are cross-polarized (turned 90° with respect to each other). You can see that no light can pass through the two films (the dark patch around the crystals). However, the lower mineral allows light to pass through the films, whereas the upper mineral does not. (*Photos by E. K. Peters*)

second film on the first, and rotate just one of them slowly until the world around you goes black (or nearly so). At this point, the two films are like two coin slots turned to lie at 90° to each other, so no light waves can pass through the two together. In other words, light waves that pass through the first film are all polarized in one direction. When a second polarizing filter is placed at 90° to the first, no light can pass through (Figure 3.4A).

B3. If we interpose a mineral between the two polarizing films, you may be surprised at what you see (Figure 3.4B).

Q3.23. Interpose each mineral (**T, U, V, W, X**) between the two polarizing films. Rotate the films. Is the light that passes through each sample blocked when the two films are at 90° to each other? Complete Table 3.11 on your Answer Sheet.

B4. Recall that the passage of light through a mineral is shaped by the atom-by-atom structure within the crystal.

Q3.24. If a crystal is placed between two polarizing films turned at 90° to each other and light passes through to reach your eye, what does this mean about the passage of the light waves through the crystal?

B5. Repeat the experiment, interposing the small vial of water between the two polarized films.

Q3.25. Is the light blocked when the two films are at 90° to each other?

Q3.26. Why might the water molecules not rotate the light waves in the manner that some of the minerals do?

Part C   Double Refraction

One common mineral has two separate angles of refraction rather than one. We call this **double refraction.** This unusual characteristic is created by a complex atomic structure. Double refraction is easy to observe; it looks like a double image.

Materials

• clear minerals labeled **T, U, V, W, X**

Procedure

C1. Draw a small dot on a piece of paper. One by one, place each crystal (**T, U, V, W, X**) on top of the dot.

Q3.27. Which of the crystals has double refraction?

Part D   Directional Color Change

Materials

• tiny cube of the mineral cordierite

Procedure

D1. You have been given a cube of an unusual mineral named cordierite (Figure 3.5). Pick any face on the cube and, with that side facing toward you, hold the mineral up to the light.

Figure 3.5   Two identical cubes of cordierite at different orientations to the camera. Note the color difference. (*Photo by E. K. Peters*)

Q3.28. Describe the color you see.

D2. Now turn the cube 90°.

Q3.29. Describe the color you observe in this second orientation.

Applying What You Just Experienced

Q3.30. The highly ordered arrangement of atoms in crystals affects the passage of what through them?

Q3.31. All crystals bend light that enters them at an angle. What is the technical word for bending light?

Q3.32. Very few minerals change color due to internal absorption of light along different directions. Name one that does.

Cubic crystals, like your sample **T** (halite), will bend light (refract it), but they never change the orientation of the light waves (rotate them). All crystals other than cubic both bend (refract) and turn (rotate) light waves.

Geologists use these optical properties of crystals as an important aid in identifying minerals. Because minerals within rocks often are quite small, geologists use microscopes with built-in polarizing films to help them distinguish one mineral from another.

Lab 3.7   Interfacial Angles of Crystals

Materials

• well-formed crystal of quartz

• contact goniometer (or protractor with movable arm)

Note that the crystal is longer than it is wide. It grew this way. The crystal in your hand is showing its normal crystal form (not cleavage!). Although the particular shape of individual quartz crystals may vary, the highly ordered structure of the quartz crystal dictates that the *same angles must occur between the faces* of the quartz crystal.

In this experiment, you will use a simple instrument to measure these angles. The instrument is called a **contact goniometer** because you place the blades of the goniometer in contact with the crystal faces to measure the angles between them.

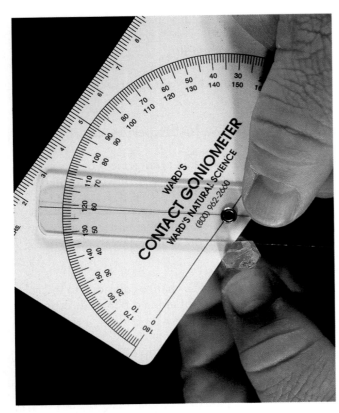

Figure 3.6 Measuring the angle between faces of a hexagonal quartz crystal with a contact goniometer. (*Photo by E. K. Peters*)

Procedure

1. Measure the angles between adjacent faces of the quartz crystal. (Don't use the pointed end of the crystal—use the faces on the long sides.) There are six sides, so you need to make six measurements around the crystal. The measuring technique is shown in Figure 3.6.

Q3.33. Record your measurements in Table 3.12 on your Lab Answer Sheet.

Q3.34. Your six measurements may not be exactly equal. Briefly, discuss possible sources of error in your measurements. (**Hint:** Consider interpolation, irregularities in the sample, and anything else that may have contributed.)

Applying What You Just Experienced

Note that these angles all have the same, or nearly the same, value. This demonstrates the **law of constant interfacial angles.** This constancy is explained by the highly ordered structure of the quartz crystal.

Crystal that grew evenly

Crystal that grew unevenly

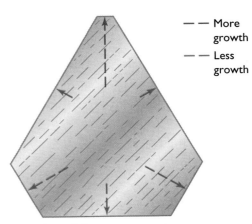

— — More growth

— — Less growth

Crystal that grew unevenly

Figure 3.7 Uneven growth of crystal faces. The smaller faces are formed by more rapid growth in their direction.

Although the angles between faces are constant, you can see that your crystal does not have six faces of exactly equal size. This is because different faces have uneven growth rates. This makes the faces unequal in size, as shown in Figure 3.7. However, the angle between any two faces always is the same because of the highly ordered atomic structure of the quartz crystal framework, or lattice.

Lab 3.8 Identifying Unknown Minerals

It's essential that you be able to identify common minerals and know their names. This means that you need to learn basic mineral characteristics and use these characteristics to successfully identify 20 to 25 common minerals. Your textbook presents important minerals and their properties. Learn the names, compositions, and characteristics of these minerals, and study their photos. Visiting mineral displays available at or near your school will help a lot.

Materials

- 20 to 25 unknown mineral specimens
- Table 3.3, Mineral Identification Flowchart
- Common mineral-testing tools

Procedure

1. In most cases, you must use more than one property to identify minerals. For example, you may use hard-

ness, cleavage, and luster to identify a specimen. (See the minerals chapter in your textbook for help.)

2. Now use Table 3.3, Mineral Identification Flowchart, to help identify your samples.

Q3.35. Complete Table 3.13 on your Lab Answer Sheet. Fill in the cells to describe your mineral specimens, and then identify them. (**Note:** Your instructor may provide more than one sample of the same mineral, and the samples might look different. Your instructor also might give you a sample that is not on the Mineral Identification Flowchart, but that will be announced in your Lab.)

Table 3.3

Mineral identification flowchart

Step 1. Does the mineral have a metallic luster?	
If YES: Go to Step 2A, Metallic minerals	**If NO:** Go to Step 2B, Nonmetallic minerals

Step 2A. Metallic minerals

Check both hardness and streak. Find the best fit, considering all features.

Hardness	Streak	Color	Other properties	Mineral name and chemical formula
6–6.5	Gray	Brassy yellow	May be tarnished; no cleavage; density around 5 g/cm^3	Pyrite (aka fool's gold) FeS_2
6	Gray	Gray to black	No clear cleavage; attracts a magnet; density around 5 g/cm^3	Magnetite Fe_3O_4
5–6.5	Reddish brown or reddish gray	Silver-gray or black	Tarnish may be reddish; streak is characteristic; not magnetic; density around 5 g/cm^3	Hematite Fe_2O_3
5–5.5	Yellow-brown	Yellow-brown to brown-black	Amorphous (no clear crystal shapes); density variable	Limonite $Fe_2O_3 \cdot nH_2O$
3.5–4	Gray	Rich yellow; tarnish may be abundant and purplish	No cleavage; softer than pyrite and darker yellow; density near 4 g/cm^3	Chalcopyrite $CuFeS_2$

Hardness	Streak	Color	Other properties	Mineral name and chemical formula
3.5–4	White to brown-yellow	Variable	Luster only barely metallic; cleavage good in six directions at 120°; density near 4 g/cm^3	Sphalerite ZnS
2.5–3	Copper color	Copper or brownish; tarnishes black-green	Malleable (bendable) and dense, 8–9 g/cm^3	Native copper Cu
2.5	Gray	Highly lustrous gray	Cleavage good in three directions (cubic); density high—7.5 g/cm^3	Galena PbS
About 1	Gray	Dark gray	Very soft; marks paper; feels slippery or greasy; low density—2.2 g/cm^3	Graphite C

Step 2B. Nonmetallic minerals

Check both hardness and streak. Find the best fit, considering all features.

Hardness	Streak	Color	Other properties	Mineral name and chemical formula
9	Scratches streak plate	Variable	No cleavage; six-sided, barrel-shaped crystals common; blue = sapphire; red = ruby	Corundum Al_2O_3
8	Scratches streak plate	Variable	Crystals often elongate in six-sided shapes; blue = aquamarine; green = emerald	Beryl (silicate mineral of complex composition; contains beryllium, Be)
7–7.5	White	Variable	Long crystals with rounded triangular cross sections; striations on crystal faces; color may change outward within crystal	Tourmaline (silicate mineral of complex composition; contains boron, B)
7	White	Variable colors to colorless	Breaks in conchoidal shapes (seashell); vitreous (glassy) to greasy luster; grows in six-sided crystals	Quartz SiO_2
7	White	Highly variable	Light colors tend to be chalcedony; darkest colors tend to be flint; jasper may be yellow-brown-red; chalcedony often banded	Chalcedony, flint, chert, and jasper; cryptocrystalline quartz SiO_2

7	White	"Bottle green" is common; varies to black and yellow	Breaks in conchoidal shapes (seashell); may occur with pyroxene	Olivine $(Fe, Mg)_2SiO_4$
7	White	Variable; commonly dark red	No cleavage; grows in dodecahedral (12-sided) shapes (or 24 or 36 sides)	Garnet (silicate mineral of complex composition)
6	White	Variable, often white or blue-gray	Straight, parallel striations on cleavage faces; two cleavages near 90° to each other	Plagioclase feldspar $NaAlSi_3O_8$ to $CaAl_2Si_2O_8$
6	White	Variable, often pink	Small, slightly rounded "blebs," not fully parallel; two cleavages near 90°	Potassium feldspar $KAlSi_3O_8$
5.5	White	Dark green to black	Two directions of cleavage near 90°; may show striations	Pyroxene (silicate mineral with Ca, Fe, and Mg)
5.5	White to light green to dark brown	Dark green to black	Two directions of cleavage near 60° and 120°; crystals usually elongate	Amphibole (silicate mineral with Ca, Fe, and Mg)
Varies, 5.5–2	Red to brown-red	Brown-red to red	Earthy luster; streak is distinctive; may be interlaced with limonite; density fairly high, near 5 g/cm³	Hematite Fe_2O_3
Varies, 5–2	White	Green common; may be yellow, brown, or white	Dull masses common; fibrous crystals possible	Serpentine $Mg_6Si_4O_{10}(OH)_8$
Varies 5–2	Yellow-brown	Brownish yellow	Usually amorphous; may replace pyrite and therefore have its shape	Limonite $Fe_2O_3 \cdot nH_2O$
4	White	Colorless, often purple; may be blue, green, or yellow	Good cleavage in six directions (octahedral); crystals grow in cubic shapes	Fluorite CaF_2
3.5–4	White	Variable, often off-white to gray	Good cleavage in three directions (rhombohedral); weakly effervesces (fizzes) in acid test if powdered	Dolomite $CaMg(CO_3)_2$
3	White	Transparent to opaque; color highly variable	Good cleavage in three directions (rhombohedral); vigorously effervesces (fizzes) in acid test	Calcite $CaCO_3$

2.5–3	Brownish gray	Black to dark brown	Excellent cleavage into thin sheets (basal); harder than muscovite	Biotite mica (complex silicate with Fe and Mg)
2.5	White	Usually white or colorless; may be colored by impurities	Excellent cleavage at 90° in three directions (cubic); crystals also grow in cubes; tastes salty	Halite $NaCl$
About 2	Light yellow	Usually lemon yellow; can have red tints	Usually in masses (no form); conchoidal (shell-like) breakage; low density, near 2 g/cm^3; yellow color distinctive	Native sulphur S
About 2	White	White to colorless; may be tinted	Excellent cleavage into thin sheets; softer than biotite	Muscovite mica (complex silicate with K)
2	White	Colorless to white	Softness is distinctive; one good cleavage direction; low density, near 2 g/cm^3	Gypsum $CaSO_4 \cdot 2H_2O$
1–2	White	White to pale brown	Generally earthy mass; fine clay particles; good cleavage in one direction	Kaolinite $Al_4(Si_4O_{10})(OH)_8$
1	White	Highly variable	Pearly luster; feels soapy; massed (no form) or foliated (leaves)	Talc $Mg_3Si_4O_{10}(OH)_2$
1	Gray	Dark gray	Very soft; luster close to nonmetallic but at least semimetallic in some places; feels slippery or greasy; low density	Graphite C

Problems

Problem 3.1. Silicate Mineral Identification. Figure 3.8 shows a common silicate mineral in two distinct colors. Even though they look different, they are the same mineral.

Q3.36. Name the mineral on your Problem Answer Sheet.

Problem 3.2. Identifying a Magnetic Mineral. Very few minerals are magnetic and will attract a magnet. Such minerals also will deflect the magnetic needle of a compass. Figure 3.9 shows such a mineral.

Q3.37. Name the mineral.

Problem 3.3. Identifying an Unusual Mineral. The photos in Figure 3.10 on page 48 show a precious-gem mineral embedded in two rock samples. In one, the gem has grown in its characteristic shape. In the other, the same mineral has adopted a deformed shape because of intense shearing pressure. The mineral is rare and has a Mohs hardness of 9.

Q3.38. Give the mineral name.

Q3.39. Give the gem name for this mineral.

Problem 3.4. Optical Refraction in Minerals. Study the optical refraction shown in Figure 3.11 on page 48.

Q3.40. What kind of optical refraction do you observe?

Figure 3.9   Common magnetic mineral and four compasses. Note the orientation of the "S" (South) direction mark on the white ring of each compass. (*Photo by L. Davis*)

Figure 3.8   Common silicate minerals. (*Photos by E. K. Peters*)

Q3.41. Recall Lab 3.6. What is the mineral shown in the photo?

Problem 3.5. Identifying Minerals from Clues. Using the clues presented, identify each mineral on your Problem Answer Sheet.

Q3.42. This relatively common mineral comes in a variety of colors, has glassy (vitreous) luster, and has excellent octahedral cleavage. What is it?

Q3.43. This common mineral breaks in flat, shiny sheets and is a dark black-brown color. What is it?

Q3.44. This is the most common mineral, or mineral family, in Earth's crust. What is it?

Q3.45. This vitreous (glassy) mineral can scratch glass, comes in a variety of dark colors, and has no cleavage but sometimes shows a good dodecahedral crystal form. What is it?

Q3.46. This earthy mineral has a reddish or reddish brown streak. What is it?

Q3.47. This bottle-green, glassy (vitreous) mineral has no cleavage and scratches glass. What is it?

Q3.48. This common silicate mineral has cleavages at 60° and 120°. What is it?

Q3.49. This mineral has a dark gray streak, and you can scratch it with your fingernail. What is it?

Figure 3.10 **Mineral specimens.** (*Photos by E. K. Peters*)

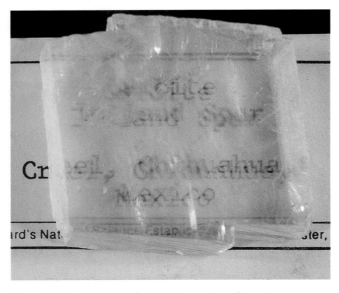

Figure 3.11 **Optical refraction in a mineral specimen.**
(*Photo by L. Davis*)

Problem 3.6. Minerals in Daily Life. We use many minerals in our daily lives either in their raw "natural" form or as part of manufactured goods.

Q3.50. In Table 3.14 on the Problem Answer Sheet, name five minerals or mineral products you use in your daily life and explain what purpose they serve.

Geo Detectives at Work   THE CASE OF THE LITTLE DARK GRAINS

Forensic geology is a special area of study. It applies geological knowledge to criminal detective work. Identifying small mineral grains at crime scenes or that cling to vehicles, clothing, corpses, or weapons can be enormously helpful to police investigations. For this reason, the Federal Bureau of Investigation (FBI) retains geologists on its staff. Smaller police organizations cannot afford to employ geologists full time, but they can send samples of Earth materials to the FBI for analysis. Similar arrangements between police and local geologists exist in other countries.

Here are two cases of forensic geology that depended on simple mineralogy. The incidents occurred in New England and Australia. Both cases hinged on small grains of a mineral called *tourmaline*.

Tourmaline (*TOUR-muh-leen*) is a silicate mineral. It contains the element boron in its crystalline structure. In large pieces, tourmaline is a semiprecious gemstone because of its deep green, red, blue, or yellow color and its hardness (7 to 7.5 on the Mohs scale). Variations in chemical composition give tourmaline its many different colors. Tourmaline's hardness helps it resist mechanical weathering, so tourmaline tends to survive as small grains in streambeds and on beaches, whereas softer minerals are abraded to microscopic dust.

In an American criminal case, a man's body was discovered on a beach. Using evidence discovered on the corpse, police concluded he had been killed at the scene. Investigators soon narrowed their focus to one suspect, but he claimed never to have been on the beach where the victim was found. He had recently walked on other beaches, however. The suspect's shoes yielded sand that was rich in grains of a black variety of tourmaline called *schorl*. Investigators noted that the beach sand at the crime scene was similarly rich in black tourmaline grains. In contrast, the beaches where the suspect claimed to have walked yielded only tiny fractions of schorl grains. The presence of black tourmaline grains in proportion similar to the site of the murder helped police tie their suspect to the crime scene.

In an Australian case, two drifters were observed traveling together. Then the nude body of one was discovered. Police questioned the other man, but he maintained his innocence. He said they had quarreled, but he left his companion alive and well at Mt. Isa, 300 miles to the east. He admitted stealing some of the other man's clothes but denied any further crime.

Police found bloodstained underwear in the suspect's possession. Sand grains adhered to the underwear, particularly the bloodstained areas. Investigators theorized that the murdered man had been wearing the article when he was killed and that the suspect had removed the dead man's clothes at the scene of the crime. They asked a geologist to compare three samples: sand from the underwear, sand from the crime scene, and sand from around Mt. Isa.

The sand at the crime scene contained many dark grains of tourmaline. Using a sophisticated lab instrument, geologists determined the chemical composition of the tourmaline, including trace elements. Although some of the sands near Mt. Isa did contain a few tourmaline grains, they had a significantly different chemical composition, or "signature." Tourmaline sand grains from the underwear proved similar to those at the crime scene. Confronted with this mineralogical evidence, the suspect confessed.

Answer the following on your Problem Answer Sheet:

Q3.51. Tourmaline belongs to what group of minerals?

Q3.52. Tourmaline is a semiprecious gemstone for what two reasons?

Q3.53. What is the cause of tourmaline's many different colors?

Q3.54. From the forensic geology viewpoint, consider the most useful type of tourmaline to find at the scene of a crime and on a suspect's clothes. Would it be the most common composition of tourmaline or one of less common composition?

Unit 3 Minerals and Gems

Last (Family) Name _____ First Name _____

Instructor's Name _____ Section _____ Date _____

Q3.1.

Table 3.4		
Solubility rates		
Sample	**Did mineral dissolve as you watched?**	**If so, how many seconds did it take?**
Halite		
Sylvite		
Gypsum		

Q3.2. (circle one)   one   three   five   can't tell

Q3.3. _____

Q3.4. _____

Q3.5.

Table 3.5				
The acid test!				
	Observation of gases (if formed)			
Mineral	**Did gas form? (yes or no)**	**What rate? (fast or slow)**	**Fizzing sound?**	**Other observations**
Calcite				
Dolomite				
Quartz				

Q3.6.

Calcite: _____

Dolomite: _____

Quartz: _____

Q3.7. _____

Q3.8. _____

Q3.9. (circle one)   yes   no

Explanation: _____

Q3.10.

Table 3.6

Rank order of mineral samples by hardness. For samples A, B, C, D, write the letter in the appropriate cell.

Letter of mineral sample	Mineral hardness
	Hardest mineral (scratches all the others)
	Next-hardest mineral (scratches minerals listed below and is scratched by the one above)
	Next-hardest mineral (scratches mineral listed below and is scratched by both minerals above)
	Least-hard/softest mineral (is scratched by all of the above and cannot scratch any of them)

Q3.11.

Table 3.7

Mohs hardness values of samples

Sample	Mohs hardness	Explain: Harder than what value? and Softer than what value?
A		
B		
C		
D		
E		
F		

Q3.12.

Diamond compared to corundum? _____ times harder

Diamond compared to talc? _____ times harder

Q3.13.

Sketch of piece 1   Sketch of piece 2   Sketch of piece 3

Q3.14.

Table 3.8

Cleavage of samples

Sample	Describe cleavage and name this cleavage type
G	
H	
I	
J	
K	
L	

Q3.15. _____

Q3.16.

Table 3.9

Hematite: mineral color vs. streak color

Sample	Describe luster (metallic? glassy? earthy? other?)	Mineral color	Streak color
M			
N			
O			
P			

Q3.17. _____

Q3.18.

Table 3.10

Density of mineral samples

Mineral	Weight (grams)	Volume (milliliters)	Density (grams per milliliter)
Q			
R			
S			

Q3.19.

1. _____

2. _____

3. _____

4. _____

Q3.20. (circle one)   yes   no

Q3.21.
Figure 3.12   Scale drawing of your crystal.

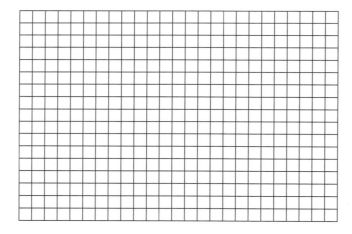

Q3.22. _____ degrees

Q3.23.

Table 3.11

Light waves in crystals between crossed polarizing sheets

Sample	Light blocked (yes or no)
T	
U	
V	
W	
X	

Q3.24.

_____

_____

Q3.25. (circle one)   yes   no

Q3.26.

_____

_____

Q3.27. Circle any of the crystals that have double refraction:

T   U   V   W   X

Q3.28.

_____

Q3.29.

_____

Q3.30. _____

Q3.31. (circle one)   diffraction   reflection   refraction   diffusion

Q3.32. _____

Q3.33.

Table 3.12

Interfacial angles of quartz crystals

Angle number	Measurement in degrees
1	
2	
3	
4	
5	
6	

Q3.34. _____

_____

_____

Q3.35.

Table 3.13

Identifying mineral specimens

Mineral sample	Luster	Mohs hardness	Cleavage	Streak color	Other notes	Mineral name
1						
2						
3						
4						
5						
6						
7						
8						
9						
10						
11						
12						
13						
14						
15						
16						
17						
18						
19						
20						
21						
22						
23						
24						
25						

Unit 3   Minerals and Gems

Last (Family) Name _____ First Name _____

Instructor's Name _____ Section _____ Date _____

Q3.36. Mineral name _____

Q3.37. Mineral name _____

Q3.38. Mineral name _____

Q3.39. Gem name _____

Q3.40. Kind of optical refraction _____

Q3.41. Mineral name _____

Q3.42. Mineral name _____

Q3.43. Mineral name _____

Q3.44. Mineral name _____

Q3.45. Mineral name _____

Q3.46. Mineral name _____

Q3.47. Mineral name _____

Q3.48. Mineral name _____

Q3.49. Mineral name _____

Q3.50.

Table 3.14

Minerals in daily life

Mineral or mineral product used in daily life	Purpose it serves
1	
2	
3	
4	
5	

Q3.51. (check one)

_____ carbonates   _____ oxides

_____ sulfides   _____ silicates

Q3.52.

1. _____

2. _____

Q3.53. _____

Q3.54. (check one)

_____ the most common composition

_____ one of the less common compositions

Molten rock underground is called **magma.** If it emerges from a fissure or a volcano onto the surface, it is called **lava.** When magma or lava cools sufficiently, it hardens to form **igneous rock.** The word *igneous* comes from an ancient word for fire, like the word *ignite.* Thus, igneous rocks are "fire-formed" (heat-formed) rocks.

Magmas can flow easily or sluggishly, depending on their **viscosity** (resistance to flow). A magma that has high viscosity has high resistance to flow and thus is sluggish, or slow-moving. Viscosity is controlled by the magma's composition, fluid content, and temperature.

When you examine an igneous rock, it is cool to the touch, of course—it has cooled from the temperature of magma down to the temperature of its surroundings. During initial cooling of a magma several processes occur simultaneously:

1. Minerals crystallize from the melt.
2. Some minerals may react with the remaining magma and re-form as different minerals.
3. The viscosity of the magma increases (it "thickens").
4. Dissolved fluids (water and gases) may separate from the melt as pressure decreases.

Each of these processes leaves a trail of evidence that is reflected in the *texture* and *mineral content* of every igneous rock. The following experiments introduce you to these concepts.

Lab 4.1 Viscosity of Liquids

Molten rock is a **liquid,** and liquids flow, no matter how viscous they are. The way magma flows within Earth and the way lava flows over Earth's surface are important to geologists who try to predict volcanic eruptions—and to fearful residents who live in volcanically active areas. In this experiment, you will work with the movement of liquids. They simulate the movement of magma and lava.

Materials
- five different liquids in small containers labeled **A–E**
- stopwatch
- thermometer
- ice
- pan or beaker
- hot plate and saucepan or hot water from tap and a container for holding it
- colored pencils (five colors)

Procedure
1. Set aside sample **E** for the moment. Study containers **A** and **B,** then **C** and **D.**

Q4.1. Can you tell that **A** and **B** are different liquids? If so, how?

Q4.2. Can you tell that **C** and **D** are different liquids? If so, how?

2. Invert a couple of the containers. Watch the air bubbles rise through the liquids. The bubbles in some of the containers rise faster than others, depending largely on the viscosity of the liquid through which they rise.

3. Your instructor will ask you to count the seconds required for the bubble in each liquid to rise to the line marked near the top of the container. A bubble reaches the line as soon as any part of the bubble first breaks the plane of the line. Practice inverting the containers and being consistent about declaring the "finish time" for each bubble.

4. Using a stopwatch or the second hand on a clock, time how long it takes for the bubble in sample **A** to rise to the "finish line" when you invert it. Do the same for samples **B** through **E.**

Q4.3. In Table 4.2 on your Lab Answer Sheet, enter your results in the "Room temperature" column.

Q4.4. Judging by your results thus far, which of the five liquids is the most viscous?

Q4.5. Which is the least viscous?

5. Keep in mind that the bonds between atoms control the properties of every substance, whether it is solid, liquid, or gas.

Q4.6. Why do you think some of the liquids are more viscous than others, and some less viscous?

6. For all the data you have recorded so far, the liquids have been at room temperature. But temperature has a strong effect on the viscosity of most liquids, including magma. Using a thermometer (or looking at the thermostat in your lab room), establish the actual 'room temperature' in your lab. Use either the Celsius or the Fahrenheit scale.

Q4.7. Record your answer at the top of the "Room temperature" column of Table 4.2 on your Lab Answer Sheet.

7. To see the effect of temperature on viscosity, hold sample **A** in a bath of ice water. (Don't submerge the container entirely; there's no need to put the cap under water.) Wait 2 minutes to let the container and its contents cool. While you wait, use a thermometer to verify that the ice-water mixture is at or near 0°C (32°F).

8. Remove sample **A** from the ice water. Measure the time required for the air bubble to rise through the sample, just as you did earlier.

Q4.8. Record your result in Table 4.2 in the "Ice-water bath" column on your Lab Answer Sheet. Do the same steps for samples **B** through **E**.

Q4.9. At the temperature of ice water, which liquid is most viscous?

Q4.10. Has the viscosity of this liquid changed because of the change in temperature?

Q4.11. Which liquid is the least viscous at ice-water temperature?

Q4.12. Does the viscosity of this liquid appear to have changed, as well as you can measure it?

9. Place your samples in a hot-water bath or hold them under running hot water in a sink (as your instructor directs). Keep them in the hot water for 2 minutes.

Q4.13. As you warm your samples, measure the temperature of the hot water and record it in Table 4.2 on your Lab Answer Sheet.

10. When your samples are warm, measure the time required for the air bubble to rise in each one.

Q4.14. Record your results in Table 4.2 on your Lab Answer Sheet.

Q4.15. At the hot-water temperature, which liquid is the most viscous?

Q4.16. How many fewer seconds does it take the air bubble in this liquid to rise to the top under hot-water conditions than it did at ice-water temperature?

Q4.17. At the warmer temperature, which liquid is the least viscous?

Q4.18. Can you see any measurable change in the viscosity of this liquid at the three different temperatures?

Q4.19. Graph your results in Figure 4.5 on your Lab Answer Sheet. First label the axes with values for seconds and temperatures. Choose scales that allow you to fit all the data on the graph. Now mark the three data points for each sample on the graph, using dots of a different color for each sample (erasable colored pencils work well). Then connect each sample's dots with a light, dashed line of the same color.

Q4.20. Judging by the slopes of the dashed lines, which sample shows the most change in viscosity from the ice-water bath to room temperature?

Q4.21. Judging by the slopes of the dashed lines, which sample shows the most change in viscosity from room temperature to the hot-water bath?

(**Note:** If you have studied calculus, you will recognize that the slope of the dashed lines on your graph roughly represents the derivative of viscosity's dependence on temperature.)

Applying What You Just Experienced

You just saw how liquids of different compositions have widely varied viscosities. So it is not surprising that magmas of different compositions also have different viscosities. This fact is extremely important in predicting the explosiveness of a volcano.

In Hawaii and Iceland, low-viscosity magmas flow out onto the surface. The lava may burn trees and houses, but the eruption is "gentle," not explosive.

In other locations, high-viscosity magmas may erupt explosively and become killers. High-viscosity magmas exist in many places, including the West Coast of North and South America. Explosive volcanoes can spread pyroclastic material for hundreds and even thousands of kilometers.

Also, each magma has a different viscosity at different temperatures. Just as in your experiment, high-temperature magmas have lower viscosities than the same magmas at cooler temperatures.

Magma viscosity also varies dramatically with the amount of silica in the molten material. Magmas with abundant silica (called felsic magma) have high viscosities (are "thicker"). Magmas with relatively low silica content (called mafic magma) have low viscosities (are "runnier"). As it happens, felsic magmas are generally low-temperature magmas, which makes them even more viscous. Mafic magmas tend to be high-temperature magmas.

Lab 4.2 Depressurization and Gas Separation in Liquids

Under high pressure, liquids can contain gases dissolved within them. An everyday example is the carbonated water used in soft drinks. Carbonated water is simply water that has carbon dioxide gas (CO_2) dissolved in it under pressure. Similarly, under high pressure, magma can have various gases dissolved in it.

A gas dissolved in a liquid will separate from the liquid if the liquid's surroundings change from a higher-pressure environment to a lower-pressure environment. Changing from higher to lower pressure is called depressurization. All magmas undergo depressurization as they rise within Earth simply because the weight of overlying rock is less toward the surface. Some magmas experience a high degree of depressurization as they erupt through Earth's surface.

Materials
- clear carbonated soft drink
- clear plastic cups
- can of shaving foam
- large container (pitcher, bucket) calibrated for volume measurements
- ruler

Procedure
1. One way to visualize depressurization in magmas and lavas is with a carbonated soft drink. As you open the soft drink, listen for the familiar hissing sound of depressurization. After you hear it, pour some of the soft drink into a cup.

Q4.22. Look carefully at the sides and bottom of the cup. What do you see?

Q4.23. The little bubbles are carbon dioxide gas. Where did they come from?

Q4.24. Where is the CO_2 in the bubbles going now?

We are done with the soft drink, so feel free to consume it. (Please rinse and save the cup for experiments to be done later on.)

2. When CO_2 gas dissolves in water, it *reacts* with the water molecules:

$$CO_2 \ + \ H_2O \ \rightarrow \ H_2CO_3$$

carbon   water   carbonic
dioxide   acid
gas

This reaction produces carbonic acid. Like all acids, carbonic acid donates a hydrogen ion (H^+) to other molecules:

$$H_2CO_3 \ \rightarrow \ HCO_3^- \ + \ H^+$$

carbonic   bicarbonate   hydrogen
acid   ion   ion

Q4.25. In light of these reactions, does a lot of CO_2 gas in the air above a city make rain droplets more acidic or more basic (the opposite of acidic) than pure H_2O would be?

3. Before you opened the soft drink, the CO_2 gas in the small air space above the liquid in the container was in equilibrium with the CO_2 dissolved in the water. CO_2 takes up more space as a gas than it does when dissolved in water.

Q4.26. Recalling the Le Châtelier principle, if you could increase the pressure in a closed container of soft drink, would more or less CO_2 gas dissolve into the water? Explain.

4. The pressure outside the soft drink bottle—the air pressure that surrounds us—is called atmospheric pressure. This pressure increases and decreases slightly with the weather, as a barometer shows us. It also decreases slightly with elevation above sea level, as an altimeter in an airplane shows us. But for our purposes, we will consider atmospheric pressure within your lab room to be steady: about 1 kilogram per square centimeter (about 15 pounds per square inch). In other words, every square centimeter of your body has about 1 kilogram of air pressure on it (every square inch has about 15 pounds). This amount of pressure is called, logically, 1 **atmosphere.**

Atmospheric pressure is great enough to force about 30 inches (about 80 centimeters) of liquid mercury metal up a narrow empty tube. This leads to another unit of pressure that you have heard weather reporters use: "inches of mercury" or "millimeters of mercury." For our purposes, 1 **bar** of pressure approximately equals 1 atmosphere.

5. Consider the can of shaving foam. Conditions inside the can are similar to those inside a soft drink container: both are pressurized. We don't know the specific amount of pressure inside the shaving foam can, but we can do a simple calculation to approximate what it might be. If you shake the shaving can, you hear that there is a little liquid and a lot of gas in the can. This gas, then, is not dissolved in the liquid. It is separate matter, what scientists call a separate "phase" from the liquid.

6. Empty the shaving foam can into a large pitcher or bucket by holding down the button until the can is fully depressurized. Note the tiny bubbles in the foam. Measure the total volume of shaving foam produced. (It's most convenient for later calculations to make this measurement in metric units, such as liters.)

Q4.27. Enter the volume you've measured in Table 4.3 on your Lab Answer Sheet.

7. How much volume did the shaving foam occupy while it still was pressurized in the can? First, we can assume that the volume of shaving foam approximately equals the volume of the can. What is this volume? How can we calculate it?

8. Note the can's cylindrical shape (Figure 4.1). It's not a perfect cylinder, because the top is a dome and the bottom of the can is depressed in a similar shape. For our purposes, the volume of the top dome of the can approximately equals the volume of the bottom depression, so they cancel each other. Thus, we can approximate the volume of the can by using the formula for the volume of a simple cylinder:

Volume of a cylinder $= \pi r^2 l$
(pi $\times$ radius squared $\times$ length)

To work the formula, follow the steps, entering the numbers in the blanks below.

a. π has a value of about 3.14. Enter this in the π blank in part d.

Figure 4.1 Cylinder.

b. Measure the can's diameter (d) in centimeters and write it here: _____ centimeters. Divide this number by 2 to get the radius (r) and write it here: _____ centimeters. Square this number and enter the result in the r^2 blank below.

c. Measure the length (l) of the shaving foam can in centimeters. (Measure just the metal part, from the bottom rolled metal edge to the top rolled metal edge.) Enter the result in the l blank below.

d. Now process the numbers to calculate the volume of the can, and write the result.

Shaving can volume $=$

$$\underbrace{\text{_____}}_{\pi} \times \underbrace{\text{_____ cm}^2}_{r^2} \times \underbrace{\text{_____ cm}}_{l} = \text{_____ cm}^3$$

e. In the metric system, 1 cubic centimeter $=$ 1 milliliter $=$ 1/1000 of a liter. Knowing this and remembering the method of unit conversion presented in Unit 1, convert the shaving foam can volume from cubic centimeters to liters.

Q4.28. Enter the result in Table 4.3 on your Lab Answer Sheet.

9. Table 4.3 is now complete except for the original pressure within the can. How can you find this value? There are instruments that can directly measure the pressure in the shaving cream can. But we can calculate a rough value by using Boyle's law. English scientist Robert Boyle (1627–1691) discovered several laws that govern the pressure and volume of gases.

10. Boyle's law applies to pure gases at constant temperatures. This is not quite the situation we are dealing with in a shaving foam can, but as an approximation, we can use this version of his law:

$$P_1 V_1 = P_2 V_2$$

where $P =$ pressure, $V =$ volume, and $_1$ and $_2$ refer to time 1 and time 2.

11. Adapting this formula to the shaving foam example, it becomes:

$$P_{1\,(\text{inside can})} \times V_{1\,(\text{inside can})} = P_{2\,(\text{outside can})} \times V_{2\,(\text{outside can})}$$

Carrying it another step:

$$P_{1\,(\text{inside can})} \times V_{1\,(\text{volume of can})} =$$
$$P_{2\,(\text{one atmosphere})} \times V_{2\,(\text{foam volume outside can})}$$

You already have measured or calculated all the factors in this equation except for $P_{1\,(\text{inside can})}$. Because you have only one unknown quantity, you can use algebra to re-arrange the equation:

$$P_{1\text{(inside can)}} = \frac{P_{2\text{ (one atmosphere)}} \times V_{2\text{ (foam volume outside can)}}}{V_{1\text{ (volume of can)}}}$$

12. Now plug in the values from Table 4.3, and solve this equation for the original pressure of the fluid in the can. Make sure your units work out, leaving you with pressure in atmospheres (or bars).

Q4.29. In Table 4.3 on your Lab Answer Sheet, write your answer for the pressure in the shaving foam can. Be sure to include the units in your answer.

Applying What You Just Experienced

Now that you have approximated the pressure in the shaving foam can, compare the result with a geologic example. Recall that 1 atmosphere of pressure equals about 1 bar. Magmas deep within Earth experience hundreds or even thousands of atmospheres, or bars, of pressure. A magma that exists about 1 kilometer beneath the surface (that's about the length of 11 football fields, straight down) can be expected to experience about 300 atmospheres, or 300 bars, of pressure from the weight of the rocks above it. Hence, magma at 10 kilometers (6 miles) beneath the surface generally exists at about 3000 atmospheres or bars of pressure.

Q4. 30. If 1000 bars equals one kilobar (kb), how many kilobars of pressure does a magma 10 kilometers beneath the surface experience?

Q4.31. How many times greater than the pressure within the shaving foam can is that?

As a rough rule, 2% to 3% of the weight of deep-underground magma is dissolved gases. Because magma at such depth may have enormous volume (many cubic kilometers), and because magma is dense and weighs a great deal, 2% to 3% of the magma's weight represents enormous quantities of gases and therefore a staggering explosive potential!

Lab 4.3 Density of Volcanic Rocks

Rhyolite and **pumice** are two common volcanic rocks. They often have similar mineral compositions. How-ever, their *texture* differs dramatically because of differences in how the lava erupts that forms each rock.

Rhyolite is a rock with a relatively "smooth" (non-porous) texture. In contrast, pumice is a rock with "fossil bubbles" (holes) that were formed by dissolved gases as they escaped from the lava during eruption. These holes are now filled with air because the volcanic gases leaked away long ago. (Geologists call these holes **vesicles** from a Latin word meaning "little bladder.") Pumice has a very porous texture with many short "edges" that makes it very abrasive. In fact, a commercial hand soap includes ground pumice as an ingredient because of its abrasive quality (Figure 4.2).

In rocks like pumice, the proportion of vesicles is truly large. This makes pumice uncommonly light in weight—in other words, its **density** (weight per unit volume) is unusually low. You are about to see just how low a rock's density can be.

The density of water is 1.0 gram per cubic centimeter (1.0 g/cm^3). Solid objects that have a density greater than 1.0 g/cm^3, such as a brick, sink in water. Solids that have a density less than 1.0 g/cm^3, such as cork, float in water.

Materials
- samples of rhyolite and pumice, roughly shaped as rectangular blocks
- container of water

Procedure
1. Place a sample of rhyolite and a sample of pumice in a container of water.

Figure 4.2 Pumice. The holes in pumice make it possible for small bits of the rock to break off fairly easily. These bits are very hard and make a good abrasive. A popular hand soap, trade-named Lava, contains ground pumice (the package says that it is "pumice-powered"). *(Photo by E. K. Peters)*

Q4.32. What can you infer about the density of each type of rock?

2. Observe the pumice sample carefully. (If its vesicles fill with water and it sinks, get a dry piece of pumice and place it in the water.)

Q4.33. What is your estimate of the proportion (%) of the pumice sample that is above the surface of the water?

Q4.34. Because the total volume of the sample is 100%, you can subtract and find the proportion of the pumice that is below the water. What is the proportion?

Q4.35. Now convert this percentage to its decimal equivalent.

Applying What You Just Experienced

Rhyolite and all other common igneous rocks on Earth have densities more than twice that of water (1.0 g/cm^3). This is not surprising when you consider what rocks are made of. With the exception of oxygen, rocks are composed principally of heavy atoms: silicon, aluminum, and iron.

Q4.36. Silicon, aluminum, and iron are relatively heavy elements compared to those that compose water. Of which two much lighter elements is water composed?

Because the solid rocks of Earth are denser than water, our streams, lakes, and oceans are kept free of floating rocks!

Q4.37. Pumice is an exception to this rule. It floats because of the numerous holes in it, and because the holes are isolated, so water cannot flow into most of them. What do geologists call these holes?

Q4.38. What material separates from liquid lava, thus forming the holes?

Ice and water have the same composition, of course—H_2O. But ice floats on water. This tells us that ice is *less dense* than water. But how can ice be less dense than water, when most solids sink in a liquid of the same composition? The answer is that ice has a crystalline structure that *takes up more space* than the same number of molecules of liquid water. This makes ice float, guaranteeing that partly frozen lakes have ice on top in the winter rather than on the bottom.

Lab 4.4 Identifying Igneous Rocks

In this lab, you will learn how to identify igneous rocks from their texture and color.

Materials
• igneous rock specimens
• Table 4.1, Igneous Rock Identification

Procedure
Use the descriptions of igneous rocks in your textbook and Table 4.1.

1. Examine each sample. In the table, determine its "Texture category" from the first column.

2. Go to the corresponding "Igneous rock category" in the second column and determine which description best fits the sample. Assign a name to your sample.
Example: If your sample has a glassy texture, it fits into texture category 1, which sends you to the igneous rock category "Volcanic." If your sample is black, then it probably is the rock called obsidian.

Q4.39. Complete Table 4.4 on your Lab Answer Sheet. Fill in the cells to describe the igneous rock specimens, and then identify them in the "Rock name" column.

Problems

Problem 4.1. Shapes of Volcanoes. Figure 4.3 on page 69 shows three volcanoes. Observe their shapes.

The slope of a hill or mountain is the measurement of how far it is inclined upward from the horizontal, expressed in degrees (°). Using a protractor, measure the slope of each volcano about a third of the way down from its top. Then measure the slope of each volcano near its base. For each measurement, choose a place where the slope looks consistent (not where it is obviously changing).

Q4.40. Record your measurements on the Problem Answer Sheet.

Q4.41. From this information alone, do you think the lava that flows from each volcano is largely silica-rich (felsic) or largely silica-poor (mafic)?

Q4.42. For each volcano, would you expect an explosively violent eruption or a more gentle one?

Table 4.1

Igneous rock identification

Step 1: Texture category		Step 2: Igneous rock category
1. Sample has *glassy* texture and maybe a conchoidal fracture.	→	**Volcanic.** *Your rock had a relatively fast cooling history and may be* a. **Obsidian:** Black or dark red or a combination of the two; conchoidal fracture. b. **Pumice:** Light-colored (usually gray); high percentage of vesicles and glass slivers; noticeably low density.
2. Sample has *aphanitic* texture (essentially a single color with few observable mineral crystals, if any). Overall single color may be shades of gray, green, pink, or black.	→	**Volcanic.** *Your rock had a relatively fast cooling history and may be* a. **Rhyolite:** Light-colored (generally light gray to pink), with some glassy slivers. b. **Andesite:** Dark to greenish gray, green, or reddish and may have tiny but visible minerals (often plagioclase feldspar). c. **Basalt:** Very dark gray to black; may contain small olivine or pyroxene or feldspar crystals; may contain a few vesicles; weathered surface appears "rusted." d. **Vesicular basalt:** Numerous vesicles
3. Sample has *aphanitic-porphyritic* texture (essentially a single color with numerous visible crystals). Crystals often are very square and usually lighter or darker than the surrounding material. The overall single color may be shades of gray, green, pink, or black.	→	**Volcanic.** *Your rock had a relatively fast cooling history and may be* a. **Porphyritic rhyolite:** Light-colored (generally light gray to pink), but with large crystals (phenocrysts) (>3 mm). b. **Porphyritic andesite:** Dark to greenish gray, green, or reddish, with large crystals called phenocrysts (>3 mm). c. **Porphyritic basalt:** Black or very dark gray, with large crystals called phenocrysts (>3 mm) that generally are lighter in color than the surrounding matrix.
4. Sample has *pyroclastic* texture (essentially a single color–usually shades of gray or black) with numerous vesicles (holes) throughout.	→	**Volcanic.** *Your rock had a relatively fast cooling history and may be* a. **Pumice:** Light-colored (usually gray) with high percentage of vesicles and glassy slivers; noticeably low density. b. **Scoria:** Dark-colored (usually black or very dark red) with high percentage of vesicles.

		c. **Tuff:** Generally light-colored (usually beige to light gray), often friable (crumbly), with high percentage of volcanic ash; may have visibly larger pieces; may have high percentage of vesicles.
5. Sample has *phaneritic* texture: you can see one or more minerals of relatively uniform size.	→	**Plutonic.** *Your rock cooled relatively slowly and may be* a. **Granite:** 65–80% light-colored minerals. Primary minerals are quartz, more orthoclase than plagioclase feldspar, muscovite or biotite mica, and amphibole. b. **Diorite:** About 50/50 light-colored and dark-colored minerals (salt-and-pepper look). Light-colored include plagioclase and orthoclase feldspars and very little or no quartz. Dark-colored include biotite mica, amphibole, and pyroxene. c. **Gabbro:** 65–85% dark-colored minerals. Primary minerals are plagioclase feldspar, amphibole, pyroxene, and olivine. d. **Peridotite:** Very dark-colored (dark green to black). Primary minerals are olivine and pyroxene.
6. Sample has a *phaneritic-porphyritic* texture, meaning that you can see two or more minerals, but the crystal size distribution is bimodal.	→	**Plutonic.** *Your rock had a complex cooling history and may be* a. **Porphyritic granite:** 65–80% light-colored minerals. Primary minerals are quartz, more orthoclase than plagioclase feldspar, muscovite or biotite mica, and amphiboles. Bimodal crystal size distribution; feldspars generally larger than other minerals. b. **Porphyritic diorite:** Both light-colored and dark-colored minerals; bimodal crystal size distribution. Light-colored include plagioclase and orthoclase feldspar; generally larger; quartz is minor or absent. Dark-colored minerals include biotite mica, amphibole, and pyroxene; generally smaller. c. **Porphyritic gabbro:** 65–85% dark-colored minerals. Primary minerals are plagioclase feldspar, pyroxene, amphibole, and olivine; bimodal size distribution.

A

B

C

Figure 4.3 **A. Nganruehoe volcano on the North Island of New Zealand.** *(Photo by W. B. Hamblin, courtesy of the U.S.Geological Survey)* **B. Mauna Loa volcano on Hawaii (the "Big Island").** *(Photo courtesy of the U.S. Geological Survey/Hawaiian Volcanoes Observatory)* **C. South side of Mt. St. Helens, September 1988.** Author Larry Davis lends a sense of scale.

Q4.43. Which volcanoes have higher-temperature magma or lower-temperature magma? (See your textbook.)

Problem 4.2. Types of Lava. Study the lava flow in Figure 4.4 on the following page.

Q4.44. Do you think the lava is felsic or mafic?

Q4.45. When the lava cools, what volcanic rock is likely to form?

Problem 4.3. Analyzing a Plutonic Rock Containing Olivine Crystals. Imagine examining a plutonic rock in which you see olivine crystals.

Q4.46. Would you expect to find quartz as well? Why? (***Hint:*** Think about Bowen's reaction series or the process of cooling and crystallization.)

Problem 4.4. Analyzing a Plutonic Rock Containing Muscovite. Imagine examining a plutonic rock in which you see muscovite.

Q4.47. Would you expect to find olivine in the rock as well? Why?

Problem 4.5. Identifying Igneous Rocks. Imagine that you are the geoscience officer on a starship. You travel to a galaxy far away, where the ship visits an Earth-size planet. The minerals on this planet are quite different from the silica-based minerals you know on Earth. You therefore name the planet Xenon, from the Greek word meaning "strange" or "foreign."

On Xenon, there is a pink mineral that has a high melting temperature. You experiment and find that, as you cool the magma, the pink mineral reacts with the molten material and forms a yellow mineral. Over time, all the pink mineral crystals disappear and yellow crystals are formed. The yellow mineral, in turn, reacts with magma as the temperature drops farther and is replaced by a green mineral.

Q4.48. Consider Bowen's reaction series or the process of cooling and crystallization. The pink mineral on Xenon is like what Earth mineral?

Q4.49. The yellow mineral on Xenon is like what Earth mineral?

Q4.50. The green mineral on Xenon is like what Earth mineral?

Q4.51. If a rock on Xenon has uniformly sized large crystals, what can we say about the cooling history of the rock?

Problem 4.6. Practical Application of Igneous Rock Knowledge. Foundries make molds into which molten metal is poured to cast objects. Sometimes the molds are made of compacted olivine sands. The sands are compressed, a mold is made in the sand, and the metal is poured into the mold. When the metal cools, the object is knocked out of the loose sand, which is then reused repeatedly to make fresh molds.

Q4.52. Why do foundries go to the trouble and expense of making their molds out of olivine sands rather than much cheaper quartz sand? (***Hint:*** Consider Bowen's reaction series or the process of crystallization.)

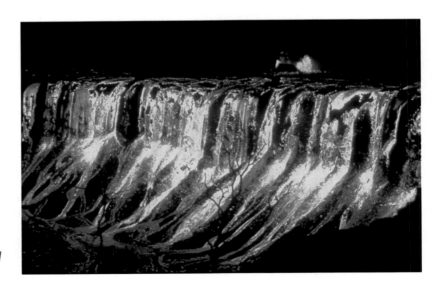

Figure 4.4 Kilauea volcano, December 1969. *(Photo by J. B. Judd, courtesy of U.S. Geological Survey)*

Geo Detectives at Work   THE CASE OF THE MYSTERIOUS BALLOON BOMBS

Today, we worry about incoming missiles with nuclear warheads. But in 1944, intercontinental missiles were still being researched and the first nuclear bomb was still under construction in the government's top-secret Manhattan Project. The United States was in the depths of a "conventional" war, one fought on the ground with soldiers and tanks and from the air with bombs dropped from planes. It was World War II, fought against Germany, Italy, and Japan during 1939–1945.

Although U.S. soldiers fought in many foreign lands during the war, soldiers from the other side never fought on North American soil. Nor did their planes ever reach the North American continent. However, aerial bombs did reach the U.S. and Canadian West Coast by other means.

During 1944 and 1945, about 300 incendiary bombs exploded, literally out of the blue, across Alaska, Canada, and the western United States. The bombs drifted in from the Pacific Ocean, carried by hydrogen-filled balloons. The balloons were weighted with paper bags of sand for ballast, to keep them from rising too high. The bombs started a few fires and caused a few deaths, although damage overall was minor.

Because the bombs were blowing in from the Pacific Ocean and onto the West Coast, and because prevailing winds arrive there from the direction of Japan, the U.S. military suspected a Japanese source. But how could they prove it? And how could they find the launching location to halt the balloon-bomb attack?

U.S. geologists studied the ballast sand for clues. Its grains included several igneous minerals, including augite (*AW-jite*) and hypersthene (*HY-pers-theen*), both varieties of the mineral pyroxene. These two minerals made up much more than half of the sand. This was a valuable clue, because many of the world's beaches contain no hypersthene at all, and augite is generally a tiny fraction of beach sand. The geologists

concluded that the sand originated from unusual igneous rocks, such as those in some parts of the island arc of Japan.

The sand also contained the mineral magnetite, with unusually high concentrations of titanium in its crystal structure. Magnetite is also fairly uncommon in igneous rocks.

The sand's quartz grains were analyzed for their optical characteristics. This showed that they formed at high igneous temperatures, indicating a particular kind of lava flow as a source.

Finally, the sand included tiny seashell fragments and fossilized diatoms, both of which have a definite geography.

The researchers studied reports and maps of Japan's geology. Using their knowledge of igneous rocks and fossils, they pieced together where the balloons were coming from. The geologists told the military that they should look at two Japanese beaches. Soon word came back that the geologists had been right: the launch site was one of those two beaches, near a hydrogen-producing plant. U.S. bombers then destroyed the hydrogen plant and the balloon-bomb campaign ended.

Q4.53. What three minerals or fossils helped geologists identify the source of the sand used in the balloon ballast sacks?

Q4.54. The United States ended the war with Japan by dropping the world's first two nuclear bombs on the Japanese cities of Hiroshima and Nagasaki. These bombs were dropped from aircraft that flew over the target cities. Why didn't the United States military simply launch these nuclear bombs on balloons from the West Coast and drift them over Japan, just as the Japanese did to the United States with its balloon bombs? (**Hint:** Look for evidence in the story you just read.)

Unit 4   Igneous Rocks

Last (Family) Name_____ First Name_____

Instructor's Name_____ Section_____ Date_____

Q4.1. _____

Q4.2. _____

Q4.3, 4.7, 4.8, 4.13, 4.14.

Table 4.2

Viscosity of unknown liquids

Sample	Time required for air bubble to rise to the top of the container		
	Ice-water bath 0°C (32°F)	Room temperature _____ °	Hot-water bath _____ °
A	seconds	seconds	seconds
B	seconds	seconds	seconds
C	seconds	seconds	seconds
D	seconds	seconds	seconds
E	seconds	seconds	seconds

Q4.4. _____

Q4.5. _____

Q4.6. _____

Q4.9. _____

Q4.10. _____

Q4.11. _____

Q4.12. _____

Q4.15. _____

Q4.16. _____ seconds difference

Q4.17. _____

Q4.18. (circle one)   yes   no

Q4.19.
Figure 4.5   **Viscosity of samples A through E.**

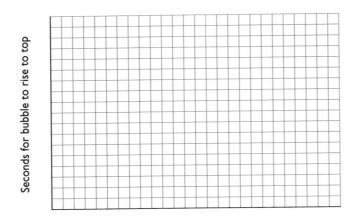

Q4.20. _____

Q4.21. _____

Q4.22. _____

Q4.23. _____

Q4.24. _____

Q4.25. (circle one)   more acidic   more basic

Q4.26. (circle one)   more   less

Explanation: _____

Q4.27, 4.28, 4.29.

Table 4.3

Depressurization and gas separation		
	Volume	**Pressure**
Shaving foam out of can (sprayed into container)	liters	About 1 bar or 1 atmosphere
Shaving foam within the can	liters	

Q4.30. _____ kb of pressure

Q4.31. _____ times greater pressure

Q4.32. The density of *pumice* is (check one)

———— greater than the density of water.

———— less than the density of water.

The density of *rhyolite* is (check one)

———— greater than the density of water.

———— less than the density of water.

Q4.33. Proportion of pumice sample above water surface: ———————— %

Q4.34. ———————— %.

Q4.35. 0.————. (The figure you have just written is the density of the pumice in grams per cubic centimeter.)

Q4.36. ———————— and ————————

Q4.37. ————————

Q4.38. ————————

Q4.39.

Table 4.4

Identifying igneous rocks

Igneous rock sample number	Visible crystals? If so, describe.	Overall texture (describe)	Silica-rich (felsic) or silica-poor (mafic)?	Rock name
1				
2				
3				
4				
5				
6				
7				
8				
9				
10				
11				
12				

Unit 4   Igneous Rocks

Last (Family) Name_____ First Name_____

Instructor's Name_____ Section_____ Date_____

Q4.40.

Figure 4.3A: Slope near top: _____ ° above horizontal

Slope near base: _____ ° above horizontal

Figure 4.3B: Slope near top: _____ ° above horizontal

Slope near base: _____ ° above horizontal

Figure 4.3C: Slope near top: _____ ° above horizontal

Slope near base: _____ ° above horizontal

Q4.41.

Figure 4.3A: (circle one)   felsic   mafic

Figure 4.3B: (circle one)   felsic   mafic

Figure 4.3C: (circle one)   felsic   mafic

Q4.42.

Figure 4.3A: (circle one)   violent   gentle

Figure 4.3B: (circle one)   violent   gentle

Figure 4.3C: (circle one)   violent   gentle

Q4.43.

Figure 4.3A: (circle one)   high-temperature   low-temperature

Figure 4.3B: (circle one)   high-temperature   low-temperature

Figure 4.3C: (circle one)   high-temperature   low-temperature

Q4.44. (circle one)   felsic   mafic   can't tell

Q4.45. (circle one)   rhyolite   basalt   pumice

Q4.46. (circle one)   yes   no   too little info

Explanation: _____

Q4.47. (circle one) yes no too little info

Explanation: _____

Q4.48. _____

Q4.49. _____

Q4.50. _____

Q4.51. _____

Q4.52. _____

Q4.53. Three minerals or fossils that helped geologists identify the source of the sand were _____ ,

_____ , and _____ .

Q4.54. _____

_____

Sedimentary Rocks: Formed from Sediments in Water and Wind

Most of us live on a continent. Therefore, we live on the surface of Earth's continental crust, where the most common rocks are **sedimentary rocks.** They are common at the surface and beneath it to depths of several thousand feet in many places. In fact, over your lifetime, you probably will see more sedimentary rocks than igneous and metamorphic rocks combined.

The importance of sedimentary rocks can't be overemphasized. We not only live on top of them; we also use them extensively in daily life. Here is the evidence: sedimentary rocks include coal (which is burned to generate more than half of the electricity used in the United States), iron ore (used to make steel), sandstone (used as a building stone), rock gypsum (used to make the sheets of building material known as "drywall"), and limestone (used to make cement and fertilizer). Some sedimentary rocks also contain fossils of plants and animals, which provide geologists with a valuable "recording" of life's history on Earth.

Lab 5.1 Maturity of Stream Sediments

We use the term *maturity* to describe where a person is along the path from birth to old age. Similarly, geologists use the term **sediment maturity** to describe where sediment is on its path from its "birth" as rock fragments to its "old age" condition of wear and decomposition due to weathering.

Geologists use three factors to define the maturity of stream sediments:

- mineral composition of the sediment particles
- texture and shape of the sediment particles
- degree of sorting of the sediment particles

Here is a brief description of each factor.

Sediment Maturity Evidenced by Its Mineral Composition. In a streambed, the sediment near the stream's headwaters is different from the sediment downstream. Take the example of a stream that drains granite rocks. Around the headwaters and near outcrops, we find more particles of mafic minerals (rich in magnesium and iron but silica-poor). Downstream, the opposite is true: the sands and gravels generally are dominated by silica-rich minerals like potassium feldspar and quartz. The reason is that the silica bond is strong, so the more

silica bonds in a mineral, the more resistant it is to weathering. The more resistant sediment survives better and thus accumulates downstream.

Geologists say that sands that contain mafic (silica-poor) minerals are "compositionally immature." On the other hand, sands that are 100% quartz grains are called "compositionally mature," meaning that they are most common downstream, where the river has had plenty of time to "mature" (wear, decompose, and sort) its sediments.

Sediment Maturity Evidenced by Its Texture and Shape. Sediment particles tend to become more rounded as they are transported downstream and the river's continuing action breaks off the rough corners of the particles. Thus, geologists call sharp-edged particles "texturally immature" and call rounded ones "texturally mature." This is one aspect of texture, the structure given to a rock or sediment by the size, shape, and arrangement of the minerals within it.

Sediment Maturity Evidenced by Its Degree of Sorting. Moving water affects the *size range* of sediment particles. In general, a stream sorts the particles it carries by depositing them in order by size, largest (heaviest) first. This order reflects the *energy* of the stream.

- Upstream, toward the white-water part of the river system, the energetic water carries large and small particles alike. Sediments in the bed of such a stream consist of mixed large particles, small ones, and all sizes in between. Geologists call such sediments "poorly sorted," and they reflect immaturity in a sediment.

- Near the end of a long stream, the stream's energy is much reduced. The larger particles have already dropped out. Downstream, sand bars are composed of small particles that approach a single size (for example, ½₂ inch in diameter). Such sediments are called "well sorted" and mature.

On a frequency distribution diagram, the two extremes appear as shown in Figure 5.1 on the next page.

As you have seen, sediment maturity embraces several distinct ideas, including mineralogical composition, particle shape, and sorting. Figure 5.2 on the next page shows how a river system moves the products of weathering granite, in this case from the high slopes of the Sierra Nevada to the Pacific Ocean. Study it to see the contrast between immature upstream sediments and the mature downstream sediments.

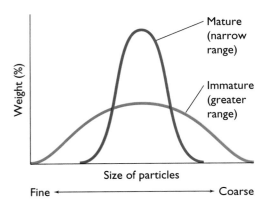

Figure 5.1 Frequency distribution showing two extreme particle-size distributions.

Materials

• For each group of three students, ten detrital sedimentary rock samples labeled **A** through **J.**

Procedure

Carefully examine the samples. Each represents a group of particles deposited at about the same time and later compacted and cemented together to form a rock.

Q5.1. Judging only by *maximum particle size,* arrange the samples in a column before you. Place the least mature sample at the "top" (farthest from you) and the most mature sample at the "bottom" (nearest to you). In Table 5.3 on the Lab Answer Sheet, record this order by size in the first column.

Q5.2. Judging only by *particle shape,* arrange your samples in a column from least mature (top) to most mature (bottom). Record this order by shape in the second column of the table.

Q5.3. Judging only by the *degree of sorting,* arrange your samples in a column from least mature (top) to most mature (bottom). Record your order by sorting in the third column of the table.

Applying What You Just Experienced

Your ordering of samples by maturity probably will vary somewhat among the three criteria. This is good, not bad. In natural science, concepts that are flexible can be highly useful—call them "shades of gray." After all, the real world is a complex place!

The concept of sediment maturity has proved useful even on other worlds. When the *Mariner* space probe was about to land on Mars, NASA worried because the probe was descending into a small valley littered with large, jagged boulders. A NASA geologist correctly inferred that the probe was above a dry streambed, full of immature sediment. Using his experience with Earth sediment samples—experience such as you are having today—he said, "Direct the probe downhill." Sure enough, the material on the ground became smaller and rounder.

Q5.4. Would a geologist call this material more mature or less mature?

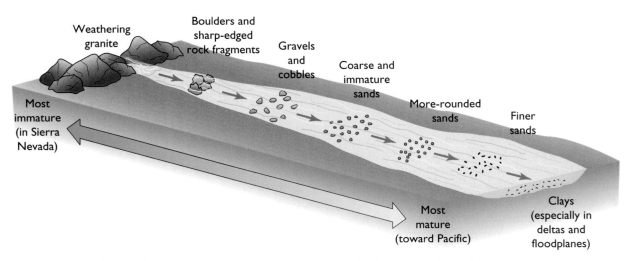

Figure 5.2 Gradation of river sediment maturity. In this example (looking generally south), maturity increases from the Sierra Nevada headwaters downstream to the Pacific Ocean.

Lab 5.2 Sieving Detrital Sediment to Sort It

Some sediment is well sorted, meaning that all the particles are the same size. The lower area of a mature river, for example, may form bars of fine, well-sorted sand. White-water rivers, on the other hand, create poorly sorted sediments with mixed particles that range widely in size. In this exercise, you will determine the particle-size distributions of several sediment samples.

Materials
- four samples of detrital sediment labeled **A** through **D**
- set of sieves
- ruler
- mass balances
- optional: hand lens for examining fine sieves

Procedure
1. Determine the size of particles that can pass through each of your sieves. The standard way of doing this is to lay a ruler on each sieve screen and count the number of holes (empty squares) in a centimeter or millimeter. (We use metric units here because they are customary in this area of geology, even in the United States.) Align the ruler with the screen wires when measuring; do not measure diagonally.

For example, one of your finer sieves may have four holes per millimeter of horizontal distance across the sieve. Thus, the size of the holes is 0.25 millimeter across (ignoring the tiny thickness of the sieve wires).

Q5.5. Record the information for each sieve on the Lab Answer Sheet. Remember to show the sizes in metric units, millimeters or centimeters as appropriate.

2. Then nest your sieves, the coarsest on the top, with finer and finer sieves downward. Beneath the lowest sieve, place the collecting pan.

CAUTION POTENTIAL EQUIPMENT DAMAGE: *Never force particles through a sieve!* Doing so can distort the holes and damage the screens.

3. Weigh the entire amount of sample **A.**

Q5.6. Record your result in Table 5.4 on the Lab Answer Sheet (next-to-last column).

4. Carefully sieve sample **A** as you have been shown by your instructor (see Unit 2). Shake the sieves gently.

Your patience will be rewarded by good separation of the different sizes of the particles.

5. Take apart the sieves and place them in a row, from most coarse to least coarse.

6. From the coarsest sieve, remove all particles and place them on a weighing tray or weighing paper. Weigh this size fraction of material. *Remember to subtract the weight of the weighing tray.*

Q5.7. Record your results in Table 5.4 (second column).

7. Repeat for the other particle sizes in the rest of your sieves.

8. Add the weights of your different size fractions.

Q5.8. Record the sum in Table 5.4. To evaluate the degree of error in your work, divide the weight of the whole sample by the arithmetic sum you have just determined. Subtract 1.000 from the result and record your answer in the column labeled "Percent difference between sum of fraction weights and weight of entire sample."

9. Ask your instructor whether your degree of error is acceptable. If not, find your problem and correct it.

10. Repeat the whole procedure for samples **B, C,** and **D** (or, if your lab instructor directs, pool your data with others in your group).

Q5.9. On the four grids in Figure 5.8 on your Lab Answer Sheet, construct histograms showing the percentage (by weight) of the sample that is in each size fraction.

Q5.10. Which of your samples is most sorted? (explain if necessary)

Q5.11. Which of your samples is least sorted? (explain if necessary)

Q5.12. Consider what you know about the size and sorting of sediment in geologic systems. What environment of deposition would be most likely for each of the four samples? For each of the four samples, choose the appropriate environment of deposition from the following list. On your Lab Answer Sheet, write the number of the environment on the line next to each sample.

1. white-water portion of a stream
2. sand bar in middle to lower part of a large stream
3. delta of a major stream

Applying What You Just Experienced

Geologists infer a great deal about *past* sedimentary environments by studying how sediments are *presently* deposited. For example, sediments that have moved in a landslide on a steep slope are highly immature. They are poorly sorted and have angular fragments. As another example, wind-blown sand grains tend to be well rounded and well sorted.

Geologists can deduce the ancient environment of deposition for many sedimentary rocks. For example, they know that a rock with angular and poorly sorted particles formed in a "high-energy" or immature environment, such as the uppermost reaches of a small white-water stream or the head of an alluvial fan.

This reasoning helps illustrate a major axiom in geology: "The present is the key to the past."

Lab 5.3 Identifying Sedimentary Rocks

Your instructor will give you samples of sedimentary rocks to identify. A few igneous rocks might be included in your sample set, so be alert for nonsedimentary textures and keep in mind the lessons of the previous unit.

Q5.13. For each sample, use Table 5.1 to answer the questions in Table 5.5 on your Lab Answer Sheet.

Table 5.1

Sedimentary rock identification

Question		Sedimentary rock type	Some possible depositional environments
Can you see shells (whole or fragments) in the sample? NO ↓	Yes →	**Fossiliferous.** *Your rock may be* a. **Fossiliferous limestone:** May be any color; fizzes with dilute hydrochloric acid (HCl); fairly soft (hardness near 3); may be coquina. b. **Fossiliferous sandstone:** Shell fragments surrounded by sand particles; may be any color; shelly fragments fizz in HCl; sand also may fizz due to calcite cement.	a. Usually warm marine, continental shelf b. Similar
Can you see plant remains in the sample, and/or does it resemble charcoal? NO ↓	Yes →	**Coal family.** *Your rock may be* a. **Peat:** Brown to brown-black, soft fragments of plants, breaks apart easily in your hand. b. **Lignite:** Brown to brown-black, plant fragments usually not visible, sample breaks apart only with serious effort. c. **Coal:** Black or shiny gray; denser than lignite or peat but much less dense than most rocks; brittle; marks paper with black streaks.	a. Swampy or boggy conditions b. Similar c. Similar

Can you see or feel clastic particles in the sample? NO ↓	Yes →	**Clastic.** *Your rock may be* a. **Breccia:** Gravel or larger clasts; particles are angular (not rounded). b. **Conglomerate:** Pebbles or larger clasts; particles are rounded (not angular). c. **Sandstone or wacke:** Sand-size clasts; clasts may be mineral or rock fragments; sandstone is "clean" (no mud) whereas wacke contains both sand and significant mud. d. **Mudstone:** May be almost any color; looks and feels like hardened mud. If most clasts are silt, rock feels gritty; if most clasts are clay, rock feels smooth. If rock breaks into roughly tabular shapes, it is ***shale.***	a. Short-lived, high-energy environment or glacial deposition b. Immature or high-energy stream or high-energy beach c. Stream, beach, or dunes d. Mature stream, lake, off-shore, or low-energy marine
Do none of the above fit?	→	**Chemical or biochemical sedimentary.** *Your rock may be* a. **Rock salt:** Usually light colors; intergrown crystals; cubic cleavage; salty taste. b. **Rock gypsum:** Usually light-colored; hardness of 2, so you can scratch it with your fingernail. c. **Limestone:** Can be any color including black; hardness near 3; fizzes vigorously in HCl (consider whether sample could be chalk or oolitic limestone). d. **Dolostone:** Can be confused with limestone, but fizzes in HCl only if powdered. e. **Chert:** Usually light-colored; breaks with a shell-like (conchoidal) pattern; scratches glass; closely related rocks are flint and jasper. f. **Ironstone:** Usually dark red, brown, or black; red to orange-yellow streak; dense.	a. Evaporite basin, warm climate b. Similar c. Usually warm marine, continental shelf d. Usually converted after deposition of limestone e. Marine, low-energy oozes f. Marine, low-energy oozes

Problems

Problem 5.1. Particle Size Distribution in a Conglomerate. As you can see in Figure 5.3, conglomerate can have two particle sizes: large (pebbles, cobbles) and very small (fine sand, clays). In other words, conglomerate can have a *bimodal frequency distribution of particle sizes.*

Q5.14. In Figure 5.9 on your Problem Answer Sheet, sketch a frequency distribution histogram for what you see.

Q5.15. Why do you think this conglomerate has such a distribution of particle sizes?

Problem 5.2. Particle Size Distribution in a Sand Dune. A geologist samples sand in a desert dune, weighs the samples, and reports the weights by size shown in Table 5.2. Calculate the percent sample weight to complete the table.

Q5.16. Using Figure 5.10 on your Problem Answer Sheet, construct a histogram showing the particle size distribution by weight as a percentage of the total sample weight.

Problem 5.3. Sedimentary Structures. Figure 5.4 shows a number of sandstone beds in Zion National Park in Utah.

Q5.17. What sedimentary structure is evident in these rocks?

Q5.18. In what setting, continental or marine, was this sediment deposited millions of years ago? What is your evidence?

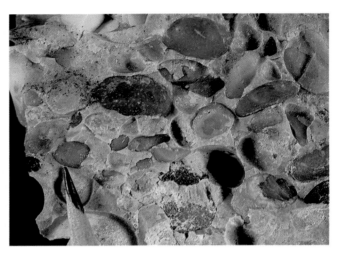

Figure 5.3 **Conglomerate sample from England.** *(Photo by L. Davis)*

Problem 5.4. Hoodoo Erosion. The two tallest columns of rock in Figure 5.5 are called "hoodoos" in southwestern Utah.

Q5.19. Consulting your textbook, what kind of erosion do you think best explains their formation? (Please supply both the name of the process and a brief explanation of what has happened here.)

Problem 5.5. Lithified Lakebed Sediments. Figure 5.6 on page 86 shows features in lithified glacial lakebed sediments.

Q5.20. What is the name of the features?

Q5.21. How much time is thought to be represented by each pair of light and dark bands?

Table 5.2

Particle size distribution in a sand dune

Particle size	Weight (grams)	Sample weight (%)
Silts and clays	20	
Fine sand	650	
Medium sand	10	
Coarse sand	(none) 0	
Pebbles and gravel	(none) 0	
Total weight	680	100

Figure 5.4   Sandstone outcrop in Zion National Park, southwestern Utah. *(Photo by E. K. Peters)*

Figure 5.5   Unusual erosion near Bryce Canyon National Park in Utah. *(Photo by E. K. Peters)*

Problem 5.6. Sandstone Cementation. Sandstone may be cemented by either silica or calcite.

Q5.22. Which type of cement is most probable in the sandstone shown in Figure 5.7?

Q5.23. Why do you think so?

Figure 5.6   **Glacial lake sediments.** *(Photo by L. Davis)*

Figure 5.7   **Eroded sandstone cliffs in Zion National Park.** *(Photo by E. K. Peters)*

Geo Detectives at Work THE CASE OF THE HEEL'S RED QUARTZ SAND

Soils are distinctive and widely varied, differing from one part of the country to another and even from one side of a hill to another. This variation is partly due to their clay mineral composition, for clay minerals compose most of the bulk material in soils. Consequently, clay minerals are often helpful in criminal investigations.

In 1906, the author of the Sherlock Holmes stories, Sir Arthur Conan Doyle, became involved in a real-life investigation. An English lawyer had been charged with killing, cutting, and mutilating animals. He was convicted and sent to prison, but Conan Doyle became convinced of his innocence. Like Sherlock Holmes himself, Conan Doyle closely observed every bit of physical evidence. He established that the lawyer's shoes—worn on the day the last crime had been committed and kept as police evidence—had black mud clinging to them. However, the field where the mutilation had occurred contained no black soils. It had only yellow, sandy clays. This simple mineralogic observation, in addition to other factors, won a pardon for the convicted lawyer.

Clay has two meanings in geology: (1) clay-*size* particles and (2) clay minerals, which are a group of minerals having various chemical compositions and different crystal structures. Mud is made of clay-size particles. Therefore, a sample of mud on the body of a hit-and-run victim, left by the car that struck the person, can be analyzed for specific clay minerals. A particular clay may exist only in a few places, especially within a given locality. Soils often are even more distinctive: they are mixtures of clays, sand, silt, and plant and animal debris called humus.

Early in the history of forensic geology, a murder case hinged on just such clues from soils. In 1908, a woman's body was discovered outside the Bavarian village of Rockenhausen. A long-time poacher with a low reputation was the chief suspect. He adamantly denied being outside town at all, insisting he was around his home all day. The police sampled soil and mud from his shoes. Just in front of the heel was a thick accumulation of mud. They sampled it in layers, from the bottommost (most recently deposited) to the topmost (first to be deposited after the suspect's wife cleaned his shoes the day before the murder.)

After examining the samples with a microscope, the police identified

- top layer: bird droppings.
- next layer: grains of red quartz sand and iron-rich clays.
- bottom layer: mud containing tiny particles of coal and brick dust.

Next, the police sampled the soils in and outside the town. They discovered

- weathered red sandstone rock at the edge of the field, where the crime occurred. The soil atop the rock contained tiny bits of the red sandstone, quartz, and iron-rich clays.
- goose droppings, thickly layered atop the soil near the suspect's house (the soil itself was composed of milk-colored quartz sand and mica grains).
- coal and brick dust atop the soil near the ruined remains of a castle outside town, a spot clearly often used by a poacher.

It now became clear that the suspect had traveled quite a bit on the day of the murder.

1. He left home in his newly cleaned shoes, acquiring the goose-dropping layer.
2. He walked to the scene of the murder, picking up tiny particles of the red-colored quartz sand and iron-rich clays.
3. He continued to what he considered a safe spot, near the ruins of the old castle, where his shoes gained their layer of mud rich in coal and brick dust.

Faced with the evidence of humble clays and other sediments on his boots, the poacher confessed and was convicted of the murder.

Q5.24. Name several different areas on an automobile or truck that would be likely to acquire coatings of clays from places where the vehicle is driven.

Q5.25. Name and describe the common soil colors on your campus (there may be several).

Unit 5   Sedimentary Rocks

Last (Family) Name_____ First Name_____

Instructor's Name_____Section_____Date_____

Q5.1, 5.2, 5.3.

Table 5.3		
Sediment maturity		
Particle size (least mature)	**Particle shape (least mature)**	**Degree of sorting (least mature)**
(most mature)	**(most mature)**	**(most mature)**

Q5.4. (circle one)   more mature   less mature

Q5.5.

Coarsest screen: _____ holes per _____ , so each hole is _____ across.

Next-coarsest screen: _____ holes per _____ , so each hole is _____ across.

Next-coarsest screen: _____ holes per _____ , so each hole is _____ across.

Next-coarsest screen: _____ holes per _____ , so each hole is _____ across.

Next-coarsest screen: _____ holes per _____ , so each hole is _____ across.

Q5.6, 5.7, 5.8.

Table 5.4

Sample data

Sample	Weight of coarsest fraction	Weight of next coarsest fraction	Weight of next-coarsest fraction	Weight of next-coarsest fraction	Weight of next-coarsest fraction	Sum of weights of all fractions	Weight of entire sample	Percent difference between sum of fraction weights and weight of entire sample
A								
B								
C								
D								

Q5.9.

Figure 5.8   Histograms showing weight percentage of each size fraction.

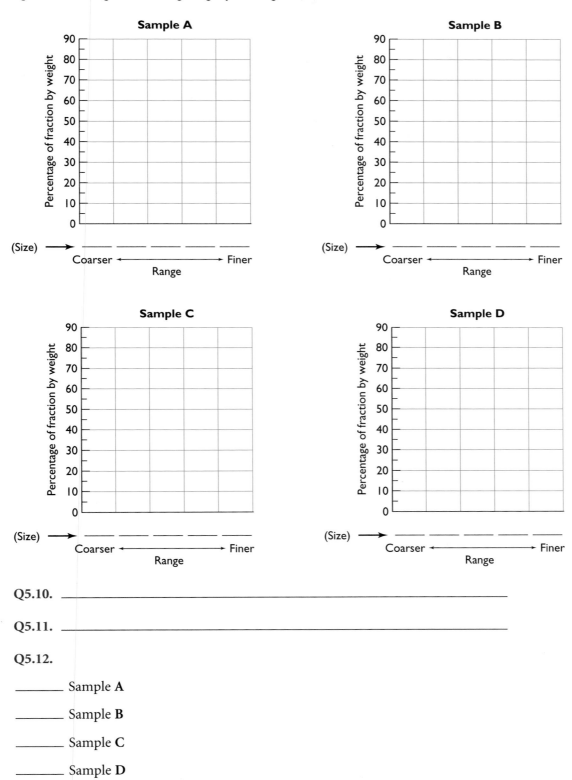

Q5.10. _____

Q5.11. _____

Q5.12.

_____ Sample **A**

_____ Sample **B**

_____ Sample **C**

_____ Sample **D**

Q5.13.

Table 5.5

Identifying sedimentary rocks (some igneous rocks may be included)

Sample	Describe clasts, if present	Describe interlocking crystals, if present	Describe acid reaction, if appropriate	Describe rock hardness, density, or other helpful characteristics	Rock name
1					
2					
3					
4					
5					
6					
7					
8					
9					
10					
11					
12					
13					
14					
15					
16					
17					
18					
19					
20					

Unit 5   Sedimentary Rocks

Last (Family) Name_____ First Name_____

Instructor's Name_____ Section_____ Date_____

Q5.14.

Figure 5.9   Grid for drawing histogram of particle-size distribution in a conglomerate. Make a rough visual estimate and draw the histogram.

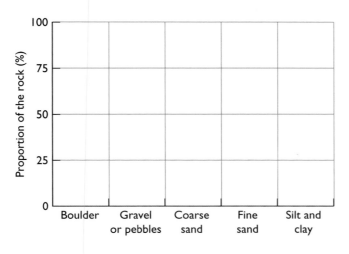

Q5.15. _____

_____

Q5.16.

Figure 5.10   Grid for drawing histogram of particle-size distribution in a sand dune by weight, as a percentage of the total sample weight.

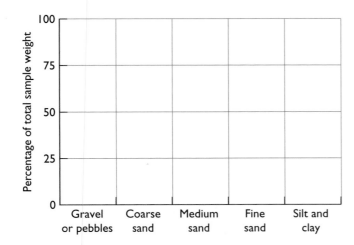

Q5.17. _____

Q5.18. (circle one)   marine   continental   can't tell

Your evidence: _____

Q5.19. _____

_____

_____

Q5.20. _____

Q5.21. _____

Q5.22. (circle one)   silica   calcite   can't tell

Q5.23. _____

Q5.24. _____

_____

Q5.25. _____

_____

Metamorphic Rocks: Heated, Pressured, Transformed

Meta means "change"; *morph* means "form." So **metamorphism** is a "change in form." If chemical and physical changes occur in a rock while it is in the solid state (not molten), and these changes alter its appearance, structure, or composition, the rock is "changed in form." Thus, the altered rock is termed **metamorphic.** The three most important physical and chemical conditions that can create metamorphic rocks are *high temperature, high pressure,* and the *circulation of chemically active fluids.* The fluids most commonly involved are water and carbon dioxide.

It's important to remember that *any* kind of rock— igneous, sedimentary, or metamorphic rock—can become metamorphosed. For example:

- Granite, an *igneous* rock, can be metamorphosed into gneiss, a metamorphic rock.

- Sandstone, a *sedimentary* rock, can be metamorphosed into quartzite, a metamorphic rock.

- Marble, already a *metamorphic* rock, can be remetamorphosed into a different metamorphic rock: another marble!

Lab 6.1 Garnet: A Classic Metamorphic Mineral

Garnet is a familiar metamorphic mineral. It often occurs in metamorphic rocks called **schists** or in the weathered sediment produced from such rocks. Garnets often have a clearly developed crystal shape (commonly dodecahedral, meaning "12 faces").

Garnets that have good, clear colors are used in jewelry. Small, industrial-grade garnets are used as an abrasive because garnet is quite hard. The garnets are crushed to sand-grain size or finer and glued to a tough paper or cloth and used as a variety of sandpaper. Garnet has a variable hardness up to 7.5 on the Mohs scale, a value slightly higher than quartz.

Materials
- 3 garnet crystals
- contact goniometer (or protractor with movable arm)

Procedure
1. Count the number of sides on each garnet.

Q6.1. How many sides does each garnet have? Record your answer on the Lab Answer Sheet at the end of the unit.

2. Select the garnet having the fewest crystal faces. Using a contact goniometer, measure the angle between any two *adjacent* faces on the garnet. Repeat for the other pairs of faces.

Q6.2. Record the angles you measure between the adjacent faces on the Lab Answer Sheet.

Q6.3. Recalling what you learned about minerals earlier in this course, what is the name of the rule governing the angles of such pairs of crystal faces?

Applying What You Just Experienced
Note that the *mineral* and *rock* portions of this course are fully intertwined. Geologists who study metamorphic rocks must be particularly good mineralogists.

Q6.4. Recall what you learned about minerals earlier in the course. Why do the garnet crystals have different colors?

Certain garnets are rich in magnesium and are known as **pyropes.** Geologists who study metamorphic rocks study the "partitioning," or relative concentration, of the element magnesium in pyrope garnets and pyroxene. The greater the temperature of metamorphism, the more magnesium exists in garnet compared with certain pyroxene crystals. This is an example of a "geothermometer" that can tell us the temperature at which an Earth process proceeds.

Q6.5. For this mineralogical relationship, does higher pyrope content in garnets denote higher or lower temperature?

Lab 6.2 Fluids Are Crucial for Metamorphism

As you will see in this experiment, fluids can dramatically promote chemical changes.

Materials
- small jar or vial
- baking soda

- crushed aspirin
- water

Procedure

1. Mix the soda and aspirin together in the jar.

2. Cover the jar and shake the contents well.

Q6.6. Do you observe any chemical reaction (chemical change) in the mixture? Record your answer on the Lab Answer Sheet.

3. Now add a little water to the sample—about ¼ to ½ teaspoon.

Q6.7. Describe what you see.

Applying What You Just Experienced

Picture a rock that is buried deep within Earth under high temperature and pressure. Under these conditions, the rock may not develop metamorphic minerals, or it may do so at an extremely slow rate. The reason is that, to form new minerals, the elements in the rock must be able to *move* from one existing mineral to a new mineral.

If we add a circulating chemically active fluid to the rock (most commonly water or carbon dioxide), metamorphic minerals are much more likely to form. Both water and carbon dioxide can dissolve many elements and transport them from an existing mineral to a site where a new metamorphic mineral is forming.

Metal atoms, such as copper, platinum, silver, and gold, also can be dissolved, transported, and deposited by circulating fluids. In your textbook, review the discussion of ore deposits associated with metamorphism.

Q6.8. What type or types of ore deposits depend especially on fluids and metamorphic conditions to form?

Lab 6.3 Heat Speeds Metamorphism

This experiment shows the important role of higher temperature in accelerating some chemical changes.

Materials
- small white crystals (unidentified)
- cup or beaker
- graduated cylinder
- spoon
- thermometer
- cold and hot tap water

Procedure

1. Using the thermometer, measure the temperature of the cold water.

Q6.9. Record your result in Table 6.2 on the Lab Answer Sheet.

2. Add *cold* water to the graduated cylinder until it reads 150 milliliters. Then pour the cold water into the cup or beaker.

3. Measure one level spoonful of the white crystals.

4. Using a watch or clock, measure the time (in seconds) required to add the crystals to the water and stir until fully dissolved. Repeat as necessary until you achieve repeatable results—that is, until each try takes about the same number of seconds.

Q6.10. Record your results (in seconds) in Table 6.2.

5. Using the thermometer, measure the temperature of the *hot* water.

Q6.11. Record the result in Table 6.2.

6. Fill your graduated cylinder and cup or beaker with *hot* water and let them sit for a minute, allowing the containers to warm. Then empty them.

7. Add *hot* water to the graduated cylinder until it reads 150 milliliters. Then pour the hot water into the cup or beaker.

8. Once again, measure one level spoonful of the white crystals.

9. Using a watch or clock, measure the time (in seconds) required to add the crystals to the water and stir until fully dissolved. Repeat several times as necessary until you achieve repeatable results—that is, until each try takes about the same number of seconds.

Q6.12. Record your results (in seconds) in Table 6.2.

Q6.13. How many times faster did the crystals dissolve in the hot water than in the cold water? To determine this, divide your hot water result (in seconds) into the cold water results (in seconds).

Applying What You Just Experienced

Even though metamorphism does not involve simple dissolution as in your experiment, temperature often is critical to accelerating metamorphic reactions. The higher the temperature of a mineral or rock—while still below its melting point—the more likely metamorphic minerals and textures are to form. Indeed, if temperatures are low enough, metamorphic reactions generally do not occur.

Q6.14. Consult your textbook about the role of temperature in the metamorphic process. With increasing temperature, what metamorphic rocks form in sequence from shale and slate?

Lab 6.4 Identifying Metamorphic Rocks

Your instructor will provide samples of common metamorphic rocks. Use Table 6.1 to describe and identify each sample. Rocks that are not metamorphic and that

---

Table 6.1

Metamorphic rock identification

A. Examine sample for coarseness of grains.
  1. Sample is dominated by medium to coarse grains or crystals.
      a. No preferred alignment or orientation of grains or crystals. **Go to B.**
      b. Definite preferred alignment or orientation of grains, OR minerals are segregated into distinct bands or zones, usually alternating light and dark bands. **Go to C.**
  2. Mineral grains not readily observable; usually dark colored (black, dark gray, red, green); surfaces generally smooth and flat. **Go to D.**

B. Examine sample for luster and fracture.
  1. High luster, and effervesces (fizzes) with weak hydrochloric acid (HCl): **marble.**
  2. Broken grains may have vitreous luster and conchoidal fracture; sample scratches glass: **quartzite.**

C. Sample has definite preferred alignment or orientation of grains OR minerals are segregated into distinct bands or zones.
  1. Medium to coarse grains of mica with strong preferred orientation; surface may glitter when rotated.
      a. No crystals of garnet or staurolite: **mica schist.**
      b. Small to large garnets present: **garnet schist.**
      c. Staurolite crystals present, sometimes forming crosses: **staurolite schist.**
  2. Obvious alternating bands of light-colored and dark-colored minerals (dark usually are biotite and amphibole; light usually are quartz and feldspar): **gneiss.**
  3. Medium to coarse grains, principally amphibole, with a strong preferred orientation; very dark green or black: **amphibolite.**
  4. Dark green, darker and lighter bands, texture may look "swirly"; hardness is 4–6 on the Mohs scale; has pearly luster: **serpentinite.**
  5. Dark to light green, strong foliation, harder than serpentinite with a mineral assemblage of chlorite, plagioclase feldspar, epidote, and calcite: **greenschist** or **metabasalt.**

D. Mineral grains not readily observable; usually dark colored (black, dark gray, red, green); surfaces generally smooth and flat
  1. Very dark (black to very dark gray); flat surfaces not generally observable; sample scratches glass: **hornfels.**
  2. Dull, dark (black to dark gray, red, or green), with flat, smooth surfaces; usually splits into sheets: **slate.**
  3. Dark (usually black to gray) with glossy sheen like Spandex; surfaces smooth but not always flat: **phyllite.**

you have seen and studied earlier might be included. Be alert for nonmetamorphic textures and mineralogy.

Q6.15. Record your identifications in Table 6.3 on the Lab Answer Sheet.

Problems

Problem 6.1. Phase Diagrams. Phase diagrams are helpful in studying metamorphic rocks. They have been constructed from experimental data gathered in labs. Figure 6.1 shows the phase diagram of the aluminosilicate minerals kyanite, andalusite, and sillimanite. The three minerals are **polymorphs,** meaning that they have the *same chemical compositions* but *different mineral structures.*

The figure shows the range of pressure and temperature over which each of the three minerals is stable. Notice that pressure increases downward, not upward, in the figure. Geologists invert the *y* axis like this because pressure increases downward within Earth. Indeed, depth is roughly indicated as an equivalent to pressure on the right-hand side of the figure. Using the figure as your guide, answer the following questions.

Q6.16. Which of the three polymorphic minerals is stable at 300°C and 100 bars (or 0.1 kilobar)?

Q6.17. If this mineral remained at 300°C but experienced an increase of pressure to 5000 bars (5 kilobars), what mineral would be stable?

Q6.18. Reasoning with the Le Châtelier principle, which of the two minerals you just named has the denser structure?

Q6.19. Which of the three minerals is stable at 6 kilobars and 400°C?

Q6.20. If this mineral remained at 6 kilobars of pressure but was heated to 800°C, what mineral eventually would form?

Q6.21. Using the Le Châtelier principle, would you expect this transformation to be one that *releases* heat (exothermic) or one that *absorbs* heat (endothermic)? Explain.

Problem 6.2. More than One Rock Can Become Gneiss

Q6.22. What other rock, in addition to shale, can end up as a gneiss under the right conditions of pressure and temperature?

Q6.23. Why do these two apparently different rocks both lead to the same metamorphic rock?

Problem 6.3. Metamorphic Facies. Metamorphic zones (facies) can be shown on a diagram that indicates the pressure and temperature range of each facies (Figure 6.2). Note that pressure increases downward on the figure, just as pressure increases "downward" within Earth.

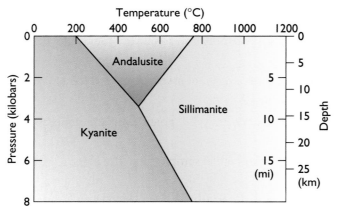

Figure 6.1   Phase diagram of three aluminosilicate minerals: kyanite, andalusite, and sillimanite. (*After M. J. Holdawayu,* American Journal of Science, *Vol. 271, p. 97.*)

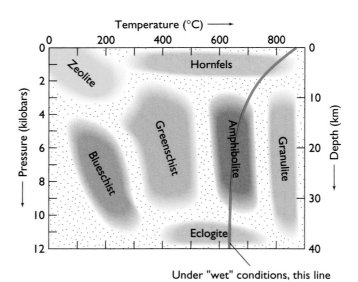

Under "wet" conditions, this line shows beginning of partial melting of rock, and therefore the end of metamorphism for "wet" rocks.

Figure 6.2   Metamorphic zones (facies) and conditions of temperature and pressure. (*From F. Press and R. Siever,* Understanding Earth, *2nd edition. New York: W. H. Freeman and Company, 1998, Fig. 8.13.*)

After studying Figure 6.2, give the metamorphic facies name that best fits each of the following sets of conditions.

Q6.24. Suppose that a mafic magma is squeezed upward, very near Earth's surface. It cools to form dikes and sills. Around the dikes and sills, which facies of metamorphism may we expect?

Q6.25. Suppose that an oceanic plate and the sediments on top of it are subducted under a relatively cool continental plate. Pressure on the down-going (subducted) rocks increases rapidly while temperature remains relatively low. This produces which facies of metamorphism?

Q6.26. Sediments in the Mississippi River settle out, building the river's delta into the Gulf of Mexico through time. As the mass of sediments increases, the pile slowly sinks. Consequently, the sediments gradually experience a modest increase in both pressure and temperature. This leads to which facies of metamorphism?

Q6.27. Folding of a mountain belt under compressional pressure gradually produces moderate metamorphism across a whole region. Most of the rocks will reflect which facies of metamorphism?

Q6.28. The highest-pressure metamorphism known to geologists is what facies?

Q6.29. The highest-temperature style of metamorphism that occurs in dry rocks at 3 to 10 kilobars of pressure is which facies?

Q6.30. What process occurs at temperatures even higher than those in your answer to Q6.29?

Problem 6.4. Using Chemical Formulas for Metamorphic Rocks. Look up the chemical formula for the minerals named on your Problem Answer Sheet. These minerals are part of some common metamorphic chemical reactions.

Q6.31. Write the formulas in the spaces provided with each mineral name. Check to see that you have a balanced chemical reaction. Then enter the missing mineral name in each reaction.

Problem 6.5. Can You Transform Gneiss Back into Shale? Increasing temperature and pressure can transform a shale into a gneiss.

Q6.32. Can you explain why gneiss is not converted back to shale as temperature and pressure decrease?

Geo Detectives at Work   BETRAYAL BY STAUROLITE

The science of **forensic geology** is a small but helpful part of some criminal investigations. Since around 1900, crime labs have analyzed soils, sands, clays, and rocks that have become evidence in criminal investigations. An interesting example comes from the 1960s.

The FBI was investigating some would-be terrorists who were believed to be moving explosives along the U.S. East Coast. At a crucial point in the FBI's investigation, a vehicle belonging to the group passed through southern New Jersey. Along the vehicle's route, agents noted large rocks at the side of the road near major intersections. This might not seem odd, except that lower New Jersey has no rock outcrops! It is composed entirely of sands and muds accumulated over eons from the ocean. Thinking that the rocks might be route markers placed by the driver, the FBI collected samples of the rocks and sought the opinion of forensic geologists.

The rocks were very distinctive, of a type known to geologists as *schist* (from a German word meaning "split"). This particular schist contained an unusual form of the mineral *staurolite*. A geologist working at the Smithsonian Institution told the FBI that the only schists containing this form of staurolite in the United States were located in western Connecticut.

The FBI followed this lead, interviewing people in western Connecticut who lived or worked near outcrops of the particular type of schist. In time, they discovered the group's store of explosives, next to an outcrop of the schist. They had used the rocks at the foot of the outcrop as markers simply because they were handy. Perhaps, while serving their prison sentences, they took time to ponder how a little geological knowledge would have helped them avoid their crucial error!

Q6.33. What features of a rock could make it a good clue as a "tracer" in criminal investigations? Briefly explain your reasoning.

Unit 6 Metamorphic Rocks

Last (Family) Name_____First Name_____

Instructor's Name_____Section_____Date_____

Q6.1.

1. _____ sides

2. _____ sides

3. _____ sides

Q6.2.

Angle between first pair of faces: _____ degrees

Angle between second pair of faces: _____ degrees

Angle between third pair of faces: _____ degrees

Q6.3. _____

Q6.4. _____

_____

Q6.5. (circle one)   higher   lower

Q6.6. (check one)

_____ I observe a chemical reaction.

_____ I do not observe a chemical reaction.

Q6.7. _____

_____

Q6.8. _____

Q6.9, 6.10, 6.11, 6.12.

Table 6.2

Relationship of time and temperature in dissolving a substance

Water	Temperature (indicate °F or °C)	Time to dissolve crystals (seconds)	Volume of water (milliliters or cubic centimeters)
Cold water			150
Hot water			150

Q6.13. How many times faster? _____

Q6.14.

shale → slate → _____ → _____ → _____

Increasing temperature →

Q6.15.

Table 6.3

Common metamorphic rocks (other rocks may be included)

Sample	Visible minerals	Texture and foliation (describe)	Rock name
1			
2			
3			
4			
5			
6			
7			
8			
9			
10			
11			
12			
13			
14			
15			

Unit 6 Metamorphic Rocks

Last (Family) Name_____ First Name_____

Instructor's Name_____ Section_____Date_____

Q6.16. _____

Q6.17. _____

Q6.18. The denser mineral is _____

Q6.19. _____

Q6.20. _____

Q6.21. (circle one)   exothermic   endothermic

Explanation: _____

_____

Q6.22. _____

Q6.23. _____

_____

_____

_____

Q6.24. _____

Q6.25. _____

Q6.26. _____

Q6.27. _____

Q6.28. _____

Q6.29. _____

Q6.30. _____

Q6.31.

Muscovite	→	Potassium feldspar	+	Water	+	

Calcite	+	Quartz	→		+	Carbon dioxide

Q6.32. _____

_____

Q6.33. _____

_____

_____

Geology's greatest contribution to intellectual history is the concept of **deep time,** truly time beyond imagining. Earth is so old compared to individual human beings—and even compared to all of human history—that we cannot help but be humbled when we contemplate our ancient planet.

There are two basic views of deep time—*relative* and *absolute.* Early geologists discovered the relationships of rocks in *relative* terms: which formed first, second, third, and so on. That is, they determined that one rock layer was older or younger than another, and that a certain fault or fold was older or younger than a rock layer.

Then, in the twentieth century, geologists learned how to date rocks in *absolute* terms—actually how many years old they are. We can now say that a particular rock is 20 million years old (approximately), or that Earth itself is about 4.6 billion years old.

In Unit 7, we explore relative time. In Unit 8, we look at ways to measure absolute time.

Geologic Time, Relatively Speaking

Geology uses two metaphors for time: the *cyclic* metaphor (events recur over and over) and the *linear* metaphor (events occur once and then fade forever into history).

- *Cyclic Metaphor.* Some aspects of geologic time appear to be cyclic, happening over and over. For example, millions of years of mountain building are followed by millions of years of erosion that reduces the mountains to flat plains; then, new uplift creates new mountains, which in turn erode to plains once again. Similarly, periods of great diversity in plants and animals are followed by mass extinctions of many species; the diminished species then diversify again to create a renewed species diversity. These processes recur over and over throughout geologic history as we know it. This is the cyclic metaphor of geologic history.

- *Linear Metaphor.* Other aspects of geologic time appear to be linear—they occur only once. When Earth formed about 4.6 billion years ago, it was so hot that it was molten. It has gradually cooled, and it continues to cool. Ultimately, Earth is expected to cool to the frigidity of surrounding space, and life and movement on Earth will cease. In this linear metaphor, there is no cycle of renewal: Earth is born, ages, and dies. (*Note:* Earth's cooling process is complex and does not occur at a linear *rate,* but it is a linear, or one-way, process as opposed to a cylic process.)

In the early days of geology, these two views of time were considered to be competing, and debates flared over which metaphor was "correct." But today, we know that both metaphors—linear and cyclic—are important. It may help you to think of Earth's history as a spiral: some parts of Earth history repeat in a cycle, over and over, but the system is linear in that it "moves forward" through time.

Initially, geologists had no clue to the **absolute** dates of Earth events. They had no idea whether Earth's birthday was 10,000 or 10,000,000 years ago. Only since about 1900 have geologists been able to put good absolute dates on rocks and minerals. For example, we can now say with some certainty that a particular rock is 11.2 million years old or 760 million years old or 3.4 billion years old. How geologists determine absolute dates is addressed in the following unit.

In this unit, we start with an easier concept, that of *relative* geologic time. Relative dating is what geologists first mastered, and you can apply it to rocks you see around your school or home. This can be intriguing, because learning to use relative time in geology is very much a matter of playing detective.

To understand relative time, let us consider for a moment several brothers and sisters. We don't know them, but judging by their height and maturity, we can put them in a relative age order: oldest sister, middle brother, youngest brother, and so on. Note that such relative ordering is quite different from saying, "this sister is $9\frac{1}{2}$ years old," which is an absolute age.

Nine logical principles underlie our ability to assign relative ages to rocks. Here is a brief summary of these principles.

1. **Original horizontality:** Most sedimentary rocks form from layers of sediment that are deposited by water or wind as horizontal layers, because gravity pulls equally on the sediment, leveling it. Therefore, many rock layers we see are parallel to Earth's surface, or "level." But if we see rock layers that are tilted or folded, we know they have been deformed at some time *after* they were deposited.

2. **Superposition:** When new layers of sedimentary or volcanic rock form, they form on top of older layers. Thus, the youngest rocks always are topmost (if they have not been disturbed). The rocks beneath become progressively older as you scan downward from top to bottom in a roadcut or look at a drill core. Because of superposition, you can determine the relative age of each rock layer. We can think of thick stacks of sedimentary rocks as representing "pages" in the "book" of Earth's history. The lowest rocks are the earliest or oldest pages.

3. **Nonconformity:** Sedimentary and volcanic rocks sometimes are deposited directly on top of existing igneous or metamorphic rocks. The contact line between the two usually is obvious and represents a gap in the geologic record, a span of time for which we don't have information.

4. **Disconformity:** Sediment is frequently deposited in shallow seas and on continental shelves. Under the proper conditions, sediment can become compacted and cemented to form sedimentary rock. If the sea level drops, these rocks become exposed to the atmosphere. Weathering erodes them until the sea level rises and deposition begins again. This forms a contact line between the old weathered layers and the new ones, a line called a *disconformity*. It is a gap in depositional history. It is an erosional surface that is roughly parallel to the bedding of the rocks. A disconformity is a useful tool in relative time study.

5. **Angular unconformity:** Sometimes we discover sedimentary rocks that have been deposited directly on top of tilted sedimentary beds. This forms a contact line that looks angular—an example of *angular unconformity*. Logically, the lower and older sedimentary beds must have been tilted and eroded before the younger sediments were deposited.

6. **Cross-cutting relationships:** Sometimes veins or dikes of igneous rock cut through other rocks. Such veins or dikes must be younger than the rocks through which they cut, like a scratch mark on a tabletop is younger than the table. Similarly, faults that break rocks must have occurred after the rocks were formed. And folds obviously are younger than the rocks that are folded.

7. **Single event horizon:** Significant ashfalls from volcanoes occasionally spread across wide areas of land and sea. The ash may be buried in a variety of depositional environments. If the ashfall is distinctive, geologists can recognize the ash layer wherever they find it. Thus it is a useful tool in relative time study.

8. **Index fossils:** Some ancient animal species had three characteristics that are valuable in relative time study:

   - They were widespread, living in major portions of the globe and perhaps in different sedimentary environments.

   - They did not survive long as a species (perhaps a million years or less, only a tiny fraction of Earth's age). Therefore, fossils of these plants and animals mark a specific time in Earth's history.

   - They had distinctive characteristics, making them easily identifiable to geologists.

   We call fossils that have these characteristics *index fossils*, because they are an index to a particular geologic time period. Geologists use index fossils to establish the relative age of sedimentary rocks worldwide.

9. **Fossil succession:** Geologists have studied undisturbed sedimentary rocks at thousands of locations worldwide. They see a clear pattern of relative ordering of species through time. All extremely ancient species were sea-dwelling single-celled organisms. Then fossil arthropods and fish appear in the rock layers, followed by fossil amphibians and then reptiles. Only more recently do dinosaur fossils appear. From these patterns it is clear that species *succeed*

one another in a regular pattern, so their fossils allow geologists to use the principle of fossil succession to determine their relative age.

You can apply each of these principles to the rocks shown in the photos in this book and to the rocks around you. Geologists in the 1700s and 1800s used these same basic concepts to construct a **relative time scale** for the entire Earth. This "calendar" is shown in Table 7.1. Notice that it has no absolute dates—the entire table is relative. Nevertheless, it shows the sequence of periods and events in Earth's rich history.

Lab 7.1 Changes in Trilobites through the First Four Periods of the Paleozoic Era

Any animal with hard parts is more likely to remain intact after death long enough to become buried and preserved as a fossil. This is why there are plenty of shellfish fossils, but jellyfish fossils are rare. Trilobites were among the first organisms on Earth to have **hard body parts.** In their case, the hard parts were their shells (Figure 7.1).

Table 7.1

Relative geologic time, some events in the history of life, and major extinctions (color lines)

Eon	Era	Period	Nicknames		Events
Phanerozoic (*phaneros* = visible or evident; *zoe* = life)	Cenozoic (*ceno* = recent; *zoe* = life)	Quaternary	Age of mammals	Ice age	*Homo sapiens sapiens* (us)
		Tertiary — Neogene			Whales and bats appear
		Tertiary — Paleogene			Primates appear
	Mesozoic (*meso* = middle; *zoe* = life)	Cretaceous	Age of dinosaurs	Chalk age	Dinosaurs diversify, then vanish
		Jurassic			Dinosaurs flourish
		Triassic			Dinosaurs appear
	Paleozoic (*paleo* = old; *zoe* = life)	Permian			Reptiles flourish
		Pennsylvanian			Coal-forming swamps
		Mississippian			
		Devonian		Age of fishes	Bony fishes and amphibians
		Silurian			Simple land plants
		Ordovician		Age of invertebrates	Jawless fishes
		Cambrian			Shelled animals (e.g., trilobites)
Proterozoic (*protero* = early or primitive; *zoe* = life)					Jellyfish and multicellular plants in ocean by end of this eon
Archean (*arch* = ancient)					Single-celled life in oceans (bacteria and algae)
Hadean (*Hades* = Hell)					Presumed molten state following Earth's formation

As it happens, trilobites were among the first fossils to be studied extensively by nineteenth-century geologists. These scientists recognized three major divisions in the body plan of a trilobite: the central, axial lobe and two "side" lobes that parallel it. Thus the name *tri* (three) *lobite* (lobes).

Trilobites lived in the oceans throughout the Paleozoic Era, but they were particularly abundant during the first four periods of that era: the Cambrian Period, the Ordovician Period, the Silurian Period, and the Devonian Period. This exercise therefore focuses on trilobites of the first four periods of the Paleozoic Era.

Trilobites are well known to geology students because many trilobite species are good index fossils. They have all the right characteristics:

- Dozens of trilobite species were widespread.
- Their hard parts were likely to be preserved as fossils.
- They are distinctive and recognizable.
- The species were short-lived, geologically speaking.

Consequently, geologists often can tell the age of a particular Paleozoic sedimentary rock by the type of trilobite it contains. If geologists spy a *Cerauras dentatus,* for example, they know that the rock must be middle Devonian.

Early geologists assumed that, over time, organisms would show "progress" or "advancement" in some manner. After all, the law of fossil succession is based on the fact that the smallest and simplest life forms are the earliest in the geologic record, and more complex, generally larger, organisms appear later. Thus, it was first believed that the trilobite fossil record would show "progress" or "advancement" over time. Many nonscientists still believe this today.

Materials
- photos in Figure 7.1
- scissors

Procedure
1. Cut out the 13 photos of trilobites in Figure 7.1. *Be careful to keep the identifying letter (**A** through **M**) with each photo.*

2. Study each of the photos. Look at differences in head and tail structures and at overall size differences.

3. Work in small groups or on your own, as your instructor directs. Arrange the trilobite photos along an imaginary timeline. Place "earlier" or "simpler" trilobites at the left and "later" or "more complex" ones toward the right.

4. Try grouping trilobites that have similar features— for example, long "points" trailing from their heads. Move your photos around, building a timeline with branching relationships.

5. When you have established a timeline, tell your instructor, who will inspect your photo arrangement and hear your reasoning. Your ideas will necessarily rest on the shape and size of the trilobites' hard parts because the hard parts are the only preserved part of the story of trilobite evolution.

Q7.1. Once your work is approved, draw a record of your ideas in Figure 7.8 on your Lab Answer Sheet. Represent each photo with its identification letter (**A–M**). Show your hypothesis about relationships by connecting the letters with lines. In other words, construct a timeline for trilobites.

6. When everyone has completed Figure 7.8, your instructor will explain the actual order in time of the trilobites you have studied. It may surprise you!

Q7.2. How much difference (a lot or a little) exists between your ordering of the trilobites and the true order shown by your instructor?

Applying What You Just Experienced
Trilobites were quite diverse throughout the Paleozoic Era, but no trilobite species survived the mass extinction of the Late Permian. The history of life on Earth makes it clear that new species arise (for whatever reasons) and species go extinct (for whatever reasons). Indeed, of all the animal species that have existed on Earth, more than 99% are now extinct! To put it another way, fewer than 1% of all species that ever lived have survived until today.

Trilobite history directly confronts notions about the "inevitable progress" that Western culture often assumes. The lesson from your work is that, although we might like the fossil record to exhibit "progress" at every level through time, it does not do so. Geologists don't know why particular trilobites flourished in the Cambrian and others in the Devonian. The record may not seem "sensible," but it is the true record of trilobite history on Earth.

Charles Darwin's theory of natural selection makes this last point in another way. According to Darwin, particular animals (such as a given trilobite) survive to reproduce not because they represent "progress" in some overarching manner but simply because they are best-suited to their local environments at the particular time they are alive.

Figure 7.1 **Fossil trilobites**
(Photos by L. Davis)

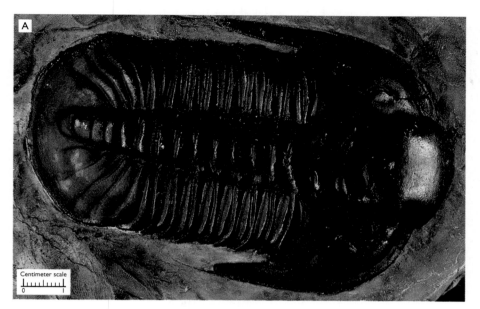

A. *Hemirhodon amplipyge* in the Wheeler Shale, House Range, Utah.

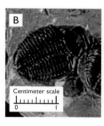
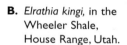

B. *Elrathia kingi,* in the Wheeler Shale, House Range, Utah.

Note: *Photo C is on the following page.*

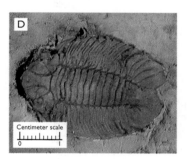
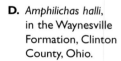

D. *Amphilichas halli,* in the Waynesville Formation, Clinton County, Ohio.

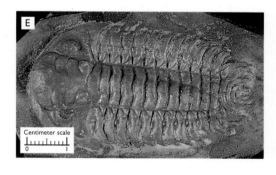

E. *Xylabion* sp., in the Bobcaygeon Formation, Ontario, Canada.

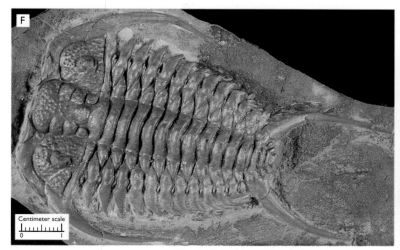

F. *Ceraurus dentatus,* in the Bobcaygeon Formation, Ontario, Canada.

G. *Diacalymene ouzregui,* Morocco, Africa.

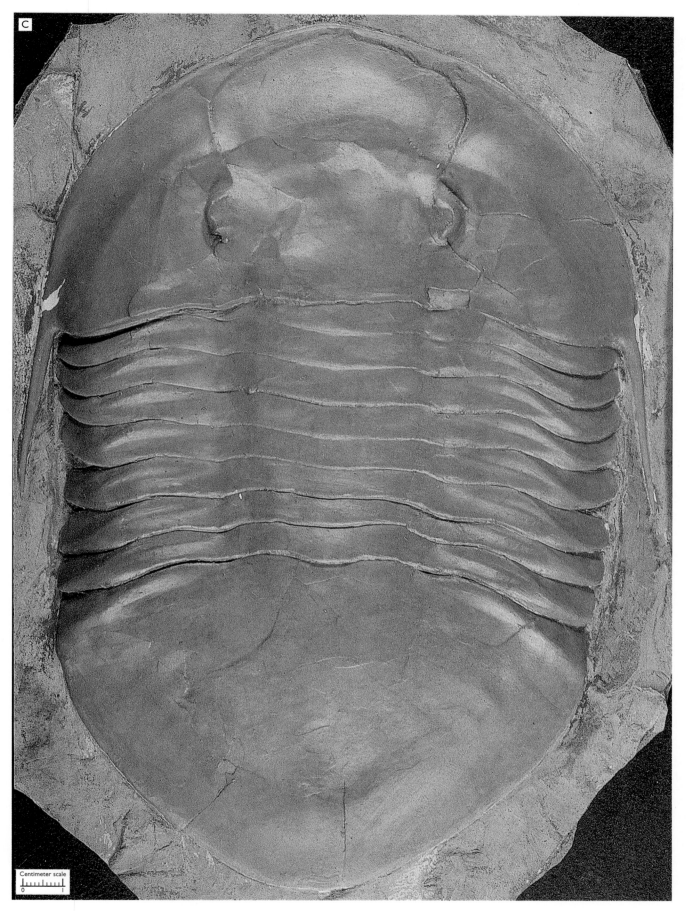

C. *Isotelus maximus,* in the Eden Shale, Cincinnati, Ohio.

Centimeter scale
0   1

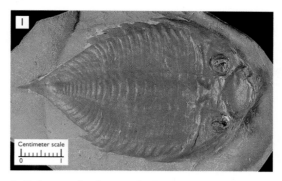

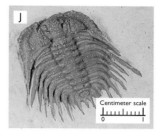

H. *Calymene celebra*, in the Edgewood Dolomite, Grafton, Illinois.

I. *Dalmanites limulurus*, in the Rochester Shale, New York.

J. *Leonaspis williamsi*, in the Haragon Formation, Oklahoma.

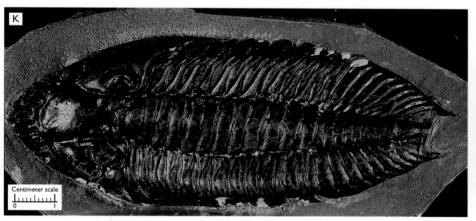

K. *Odontocephalus aegeria*, in Onondanga Limestone, Perry County, Pennsylvania.

L. *Phacops rana,* in the Silica Shale, Ohio.

M. *Phacops megalo-manicus,* Alnif, Morocco, Africa.

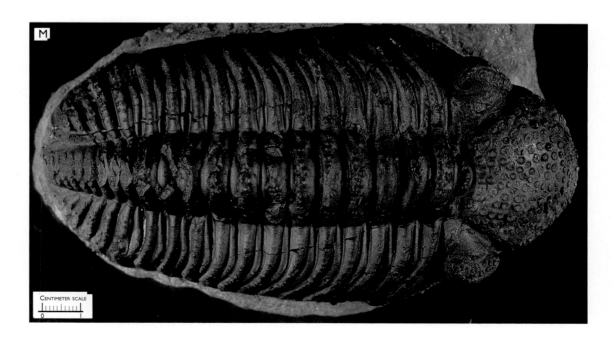

Problems

Problem 7.1. Relative Order of Events. Study Figure 7.2 to determine the relative order of geologic events that could account for what you see.

Q7.3. Record your answers in Table 7.4 on your Problem Answer Sheet. Be sure to explain fully the events that you think happened and the conditions of formation. Use at least one complete sentence for each step. (For example: "Schist formed under the high pressure and temperature of metamorphism acting on shale."). Don't skip any steps. (***Note:*** The table provides space for describing nine events, but there may have been fewer events.)

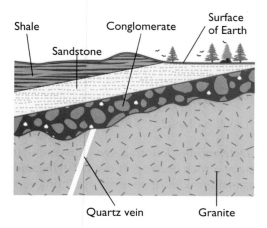

Figure 7.2   Cross section.

Problem 7.2. Relative Order of Events. Study Figure 7.3 to determine the relative order of geologic events that could account for what you see.

Q7.4. Record your answers in Table 7.5 on your Problem Answer Sheet. Be sure to explain fully the events that you think happened and the conditions of formation. Use at least one complete sentence for each step. Don't skip any steps. You may or may not need all the event spaces provided.

Problem 7.3. Relative Order of Events. Study Figure 7.4 to determine the relative order of geologic events that could account for what you see.

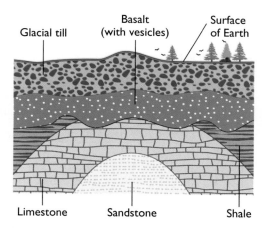

Figure 7.3   Cross section.

Q7.5. Record your answers in Table 7.6 on your Problem Answer Sheet. Be sure to explain fully the events that you think happened and their conditions of formation. Use at least one complete sentence for each step. Don't skip any steps. You may or may not need all the event spaces provided.

Problem 7.4. Cross Section Showing Relative Order of Events. Study the geological events listed in Table 7.2 on the following page. They are in order from oldest to youngest.

Q7.6. On Figure 7.9 on the Problem Answer Sheet, draw a cross section of rocks that fully represents this history. (***Note:*** Draw in pencil because mistakes are likely. You might draw a rough sketch on a separate sheet first, and then your final, neater version on the Answer Sheet.)

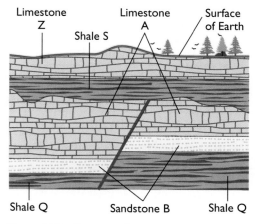

Figure 7.4   Cross section.

Table 7.2

Order of events for Problem 7.4

	Event	Description
(youngest)	8	The glacier melts, erosion occurs, and trees grow.
↑	7	A glacier deposits glacial till across the whole area.
	6	Sea level drops during the Ice Age, exposing the rocks at the surface.
	5	A shallow sea covers the area. Then sandstone is deposited, followed by shale.
	4	The gneiss is uplifted and exposed at the surface by erosion.
	3	The granite is metamorphosed to gneiss.
	2	Magma intrudes in the rocks and cools deep underground to form granite.
(oldest)	1	Rock exists deep underground.

Problem 7.5. Cross Section Showing Relative Order of Events. Study the geological events listed in Table 7.3. They are in order from oldest to youngest.

Q7.7. On Figure 7.10 on the Problem Answer Sheet, draw a cross section of rocks that fully represents this history.

Problem 7.6. Determining Relative Age. Imagine that you are looking at a rock outcrop. It is composed of coarse red sandstone that has tiny pieces of pink granite within it. A quartz vein runs vertically through the rock.

Q7.8. Which rock is oldest?

Q7.9. Which rock is youngest?

Problem 7.7. Determining Relative Age. Imagine that you are looking at a rock outcrop. It is composed mostly of granite. The granite is rimmed on top by a marble zone. This zone is overlain by limestone.

Q7.10. Which rock is oldest?

Problem 7.8. Determining Relative Age. Imagine that you are looking at a rock outcrop. It is a flat-lying and buff-colored sandstone with ripple marks pointing vertically upward. Above the sandstone is a bed of conglomerate. A thin dike of basalt cuts across both the sedimentary rocks.

Table 7.3

Order of events for problem 7.5

	Event	Description
(youngest)	8	Erosion occurs and evergreens grow.
↑	7	A new basaltic flow of lava covers the area.
	6	Volcanic ash blows into the area and settles out of the air.
	5	A basaltic lava flow covers the area.
	4	The granite is uplifted to Earth's surface as all overlying rocks are eroded.
	3	A quartz-rich vein cuts across the granite in a nearly vertical orientation.
	2	Magma intrudes in the rocks and cools deep underground to form granite.
(oldest)	1	Some kind of rocks exist deep underground.

Figure 7.5 Limestones and shales of the Green River Formation (Eocene Age), U.S. Highway 6 near Helper, Utah. *(Photo by L. Davis)*

Q7.11. Which rock is oldest?

Q7.12. Which rock is youngest?

Problem 7.9. Principles of Geologic Dating. A student reads the newspaper each day and tosses it into a box beside her desk. If her roommate says, "The oldest paper must be on the bottom," she is reasoning like a geologist.

Q7.13. Which principle of geologic dating is she citing?

Q7.14. If you grab a paper at random from the pile and look at the publication date, what kind of date are you noting?

Problem 7.10. Principles of Geologic Dating. The sedimentary rocks in Figure 7.5 are from the Green River Formation of Eocene Age in Utah.

Q7.15. What event can you say occurred in this location at some time after the Eocene?

Problem 7.11. Types of Unconformity. Figure 7.6 shows, in its lower half, the Vishnu Schist at the base of the Grand Canyon. Pink granite veins have intruded into the schist. Above the schist lies the dark Tapeats Sandstone.

Q7.16. What does the contact between the schist and the sandstone illustrate?

Problem 7.12. Types of Unconformity. Figure 7.7 on the following page shows two sets of Quaternary sedi-

ments in Idaho. The lower half of the photo shows a coarse, poorly sorted sediment. Erosion by floodwater has shaped the top of this sedimentary bed, washing away all but the larger boulders. (This layer is called a

Figure 7.6 Tapeats Sandstone overlying Vishnu Schist, Inner Gorge of the Grand Canyon. *(Photo by L. Davis)*

Figure 7.7 Lake Bonneville flood gravels underlying Lake Missoula flood sediments. *(Photo by L. Davis)*

deposit by geologists.) Above the boulders are fine and light-colored clays, intermingled with areas of darker sands (old streambeds).

Q7.17. What type of unconformity is shown between the coarse sediment and the overlying sediments?

Geo Detectives at Work   THE CASE OF THE TRULY HARD LIQUOR

Some rocks contain fossils. Geologists use fossils as evidence of when a particular rock formed during geologic time. For example, if a limestone contains a trilobite fossil, it indicates that the limestone had to have formed during the Paleozoic Era, because trilobites became extinct and do not occur in rocks of the Mesozoic or Cenozoic era. Indeed, most limestone rocks are packed with many different fossils, including shells and shell fragments. This fact is very helpful to forensic geologists.

A Canadian liquor store owner reported that he was being swindled. When he received shipments of a particular scotch whiskey, he found that some bottles had been removed from the boxes and replaced with small limestone blocks of the same weight.

The police first had to determine where the bottles were being replaced with rocks. Was it where the whiskey was distilled in Scotland or where it passed through a distribution center in England or some-

where in Canada? A geologist was engaged to examine the limestone blocks. The fossil groups in the limestone were quite distinctive. They did not occur in Canada or Scotland, but they did occur in England. Thus, the police focused their investigation on the English distributor.

Sure enough, an employee of the distributor was discovered to have access to a nearby limestone quarry. The case was clinched when witnesses stated the employee had been seen taking home small blocks of limestone from the quarry.

Q7.18. All trilobite fossils indicate that their enclosing rocks formed sometime during which era?

Q7.19. If you were a forensic geologist investigating a crime like the one described, which general type of fossil would be most helpful?

Unit 7   Relative Time

Last (Family) Name _____ First Name _____

Instructor's Name _____ Section _____ Date _____

Q7.1.

Figure 7.8   Hypothetical timeline for the trilobites.

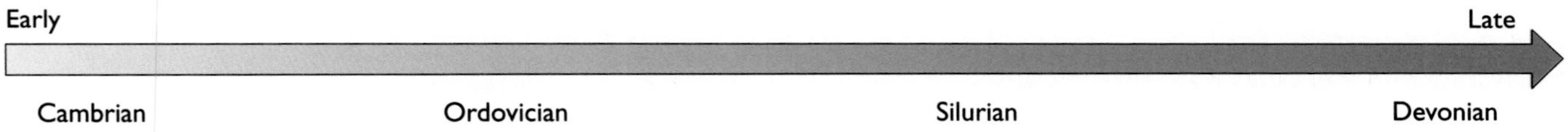

Early   Late

Cambrian   Ordovician   Silurian   Devonian

Q7.2. (circle one)   a lot   a little

Unit 7 Relative Time

Last (Family) Name —————————— First Name ——————————

Instructor's Name —————————— Section —————————— Date ——————————

Q7.3.

Table 7.4

Geologic events in relative time order (Problem 7.1)

	Event	Description
(youngest)	9	
	8	
	7	
	6	
	5	
	4	
	3	
	2	
(oldest)	1	

Q7.4.

Table 7.5

Geologic events in relative time order (Problem 7.2)

	Event	Description
(youngest)	9	
	8	
	7	
	6	
	5	
	4	
	3	
	2	
(oldest)	1	

Q7.5.

Table 7.6

Geologic events in relative time order (Problem 7.3)

	Event	Description
(youngest)	9	
	8	
	7	
	6	
	5	
	4	
	3	
	2	
(oldest)	1	

Q7.6.

Figure 7.9   Cross section.

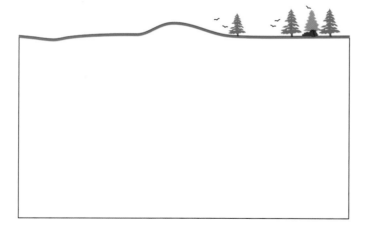

Q7.7.
Figure 7.10   Cross section.

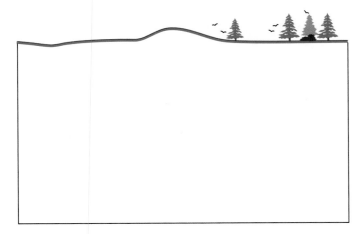

Q7.8. (check one)

_____ red sandstone

_____ pink granite

_____ quartz vein

_____ can't tell

Q7.9. (check one)

_____ red sandstone

_____ pink granite

_____ quartz vein

_____ can't tell

Q7.10. (check one)

_____ granite

_____ marble zone

_____ limestone

_____ can't tell

Q7.11. (check one)

_____ flat-lying and buff-colored sandstone

_____ bed of conglomerate

_____ thin dike of basalt

_____ can't tell

Q7.12. (check one)

_____ flat-lying and buff-colored sandstone

_____ bed of conglomerate

_____ thin dike of basalt

_____ can't tell

Q7.13. The principle of _____

Q7.14. (circle one)   relative date   absolute date

Q7.15. _____

Q7.16. _____

Q7.17. _____

Q7.18. _____

Q7.19. (check one)

_____ fossil of a distinctive organism that existed for only a brief interval of geologic time

_____ common fossil that persisted unchanged through whole eras of geologic time

Over 200 years ago, geologists learned to unravel the mystery of *relative* geologic time (Unit 7). They were able to place events in a true order—what happened first, second, and so on. They came to realize that an unconformity signifies an enormous span of missing Earth history, like whole chapters missing from a book. But how could they determine the *absolute* age of a rock layer or an event in geologic history—the actual number of years ago that something happened?

For example, they knew that flood sediments lying atop volcanic strata meant that the flood occurred after the volcano erupted, but *specifically how many years* before the flood did the volcano erupt?

In this unit, we will study the tools geologists use to determine absolute age—whether a rock is 7 million years old, 70 million years old, or 700 million years old. These tools have enabled geologists to achieve one of the great modern scientific accomplishments: accurately determining the age of Earth's oldest rocks (about 4 billion years old).

Lab 8.1   Linear Change

Rate is a change in a quantity per unit of time. Some examples of rates follow.

- 2 feet per second; 80 kilometers per hour—to indicate the rate at which you are traveling

- 175 gallons per minute; 10,000 liters per hour—to indicate the rate of water flow in a stream or through a pipe

- 14,000 short tons per day; 12,000 metric tons per day—to indicate the rate at which a mine produces coal

Rates can be linear (steady) or nonlinear (increasing or decreasing). A steady walking pace is an example of linear rate. For example, as you walk steadily across a level area, your position changes by about the same amount (feet) with each unit of time (minutes) you walk. If you walk steadily at the rate of 100 feet per minute, you will be 100 feet closer to your goal after a minute's walk, 200 feet closer after two minutes, and 50 feet closer after 30 seconds more—a linear rate. In this lab, you'll study linear rates.

Materials
- stopwatch or a watch/clock that shows seconds
- ruler or straightedge

Procedure

1. Working by yourself, go to where your instructor has made two marks labeled START and END.

2. Time how long it takes you to walk at a steady, comfortable pace from the START mark to the END mark and back again—one full lap.

Q8.1. Record this round-trip time, in seconds, for Lap 1 in Table 8.1 on your Lab Answer Sheet.

3. Using the same steady and comfortable pace, time how long it takes you to walk *two* laps.

Q8.2. Record your time in Table 8.1.

4. Now walk at the same pace for *three* laps.

Q8.3. Record your time in Table 8.1.

Q8.4. On Figure 8.5 on your Lab Answer Sheet, plot the three points that represent your times. (**Hints:** Work in erasable pencil when you plot data. If the three dots you make on the figure do not fall approximately in a straight line, reconsider your work or consult your instructor.) Using a ruler or straightedge, draw a straight line from the **origin** or "0" point through your three dots. Draw the best-fitting line you can.

5. Study your plot and answer these questions.

Q8.5. How many seconds would you need to walk a half lap?

Q8.6. How many seconds would you need to walk two and a half laps?

Applying What You Just Learned
Many geological processes have linear rates or nearly linear rates. For example, imagine that you are standing by a stream. Let's say that it flows by you at the steady rate of 1000 gallons each minute. Now, suppose a landslide suddenly blocks the stream's channel. This creates a natural dam behind which water starts to accumulate. It accumulates at about the stream's flow rate—that is, 1000 gallons per minute. Thus, after 10 minutes, 10,000 gallons of water are dammed up, and a new pond begins to form.

Q8.7. After 1 day, how many gallons of water will accumulate behind the dam?

Q8.8. After 3 weeks, how many gallons of water will accumulate behind the dam?

Lab 8.2   Exponential Change

Change does not always proceed at a linear rate. The rate of change in this experiment occurs at an exponential rate. Exponential change accelerates through time, adding onto the change already made, like a rock falling through the air, faster and faster. An example of exponential change from the business world is a savings account that earns compound interest. Its value increases at an exponential rate. Not only does the account pay interest on the principal you have deposited, it also pays interest on all the interest earned. By contrast, if you add even $1,000,000 per month to money lying in a non-interest-bearing checking account, you will see only a linear increase in the account.

Many natural processes proceed at exponential rates; a good example is the population growth of rabbits where food is abundant and no predators exist.

Materials
- box of red marbles
- box of blue marbles
- empty box
- red and blue pencils

Procedure
1. Working with your lab partner, place the three boxes in front of you: red marbles on the left, empty box in the middle, and blue marbles on the right.

2. One at a time, move all the red marbles into the empty box, *counting them as you go.* (Both you and your lab partner should count the marbles—it's important to get an accurate count.)

Q8.9. Record the number in Table 8.2 on your Lab Answer Sheet in the row labeled "Start."

3. Now, for Play 1, remove half of the red marbles and put them back in their original box. Replace *each red marble* you removed from the middle box *with a blue marble.* When you are done, you will have 50% red marbles and 50% blue marbles in the middle box—a 1:1 ratio. Record the number of red and blue marbles in the middle box in Table 8.2 at "End of Play 1."

4. For Play 2, repeat the procedure. Remove half of the remaining red marbles from the middle box, return them to their original box, and replace each red marble

that you removed from the middle box with a blue marble. Now record the number of red and blue marbles in your middle box in Table 8.2 at "End of Play 2." At this point, what is your ratio of red to blue marbles? 1:_____

5. For Play 3, repeat the procedure and record your data in Table 8.2 at "End of Play 3." At this point, what is your ratio of red to blue marbles? 1:_____

Q8.10. Now plot the results for your red marbles on Figure 8.6 on your Lab Answer Sheet. Use a red pencil. (**Hint:** If the four dots you make do not suggest a smoothly changing curve, consult your instructor.) Connect your red dots with a curving line drawn in red.

Q8.11. Now plot the results for your blue marbles on Figure 8.6. Use a blue pencil. Connect your blue dots with a curving line drawn in blue.

Q8.12. Using an ordinary black pencil, project your two curves to an imaginary fourth play. Both curves you have drawn represent nonlinear change, specifically *exponential change.* Even though the change is nonlinear, you were able to project it for an imaginary Play 4 because the change occurs at a dependable rate. You could even project the change to hypothetical Plays 5, 6, and so on.

Q8.13. Finally, using a black pencil, plot the *sum* of red and blue marbles at each "play" and at the starting point of the graph. Connect the points with a line drawn in black.

Applying What You Just Learned
For natural processes within Earth, exponential rates of change are common. A good example is the radioactive decay of unstable isotopes. Many elements that make up our planet occur in variant forms called *isotopes.* Probably best known are the isotopes of uranium. All uranium isotopes have the same number of protons and electrons, and they all behave the same way in a chemical reaction. The difference lies in the number of neutrons in the atoms of each isotope. Too many neutrons may make a uranium atom unstable, and it stabilizes itself by spontaneously radiating tiny particles or energy, a process we call *radioactive decay.* This is a nuclear reaction (a reaction within the nucleus), not a chemical reaction (a reaction between atoms).

Through radioactive decay, atoms of one isotope change themselves into atoms of another isotope. For example, after millions of years of decay, atoms belonging to one isotope of *uranium* become atoms of an isotope of the metal we know as *lead*.

The length of time required for half of a sample of an unstable isotope to change into atoms of another isotope is called the *half-life*. In the exercise you just completed, each "play" is analogous to a half-life in radioactive decay. The red marbles represented the "hot" atoms of a radioactive isotope. You made them "decay" by replacing them with blue marbles. The blue marbles represented stable or "cool" isotopes produced from the radioactive ones. Radioactive "parent" isotopes (red marbles) spontaneously decay to form "daughter" isotopes (blue marbles). The middle box of mixed red and blue marbles represented a mineral or rock that contains a mixture of such atoms.

Unlike what happened in the exercise, decay doesn't happen in short spurts scattered through time. It occurs gradually over time. Half-lives of unstable atoms range from milliseconds to millions of years. Again using uranium as an example, half of a sample of the uranium-235 isotope decays to the lead-207 isotope over a time span of about 704 million years. Thus, 704 million years is the half-life of uranium-235.

Knowing all this makes it possible for geologists to determine how long ago a mineral or rock formed. Suppose that a granite formed 1408 million years ago. And suppose that at the time it formed, it contained some unstable uranium-235. This uranium has had 1408 million years to decay through two of its 704-million-year half-lives. Thus, we expect it to have a predictable ratio of uranium-235 to lead-207. So if you find a granite that contains a ratio of 1:3 (representing two half-lives), you know that the rock must be about 1408 million years old.

Now imagine that while you are working with the red and blue marbles, your roommate enters the room. You explain what you are doing. Your roommate observes that you have 4 red and 28 blue marbles in your center box and correctly states, "You've done three plays so far, and you originally had 32 red marbles in the center box." Notice that your roommate has deduced not only the "age" of the mineral represented by the center box but the initial number of radioactive atoms in the box as well.

Let us put this in a more scientific form:

P_t = number of parent atoms (red marbles) at any time, t

D_t = number of daughter atoms (blue marbles) at any time, t

P_o = number of parent atoms (red marbles) that were initially in the center box.

Q8.14. Write a simple equation relating these three quantities, using the same reasoning your roommate did.

Q8.15. If you walked into another geology lab section and a pair of students had 8 red marbles and 120 blue marbles in their center box, how many red marbles were originally in their center box?

Q8.16. How many plays (half-lives) had the students performed with their marbles?

Problems

Problem 8.1. Determining Earth's Age from the Ocean's Saltiness. Before 1900, scientists did not know that radioactivity existed. Lacking this powerful tool for calculating the age of ancient things, geologists tried to calculate Earth's age by other means, many of which were clever. John Joly (1857–1933), an Irish professor, carefully reasoned Earth's age in this manner:

1. Rivers carry minor amounts of sodium and other components of salts into the oceans.
2. Ocean water partially evaporates into the atmosphere, forming clouds, and this evaporated water is quite pure (salt-free).
3. Therefore, over time, the saltiness of the ocean must have been created by the salts flowing in from river water and becoming concentrated by the evaporation.
4. If one could determine the average salt content of rivers and estimate the volume of river water flowing off the continents and the volume of the oceans, one could calculate Earth's age, based on the saltiness of modern oceans.

Joly focused his attention on one element in salt, namely sodium. He posited that

$$\text{age of Earth} = \frac{\text{total sodium (in oceans)}}{\text{sodium input from rivers each year}}$$

In terms of the units in the equation, convince yourself that Joly's idea makes sense (review unit analysis in Unit 1).

Joly used values for sodium in the oceans and the river volumes as understood at his time, and he reasoned that

$$\text{age of Earth} = \frac{1.418 \times 10^{16} \text{ metric tonnes of sodium (in oceans)}}{1.427 \times 10^8 \text{ metric tonnes of sodium/year}}$$

Q8.17. Calculate Earth's age as indicated by Joly's work.

Q8.18. What major assumption about salts in the ocean did Joly make that we now know to be false? (***Hint:*** Think about sedimentary rocks that form by evaporation of saltwater.)

Problem 8.2. Determining How Much Time Is Represented by Rock Strata, Based on Thickness. The nearly vertical rock face shown in Figure 8.1 is a massive sandstone in Utah's Zion National Park. Note the two rock climbers with their gear hanging on the rock face below them.

Figure 8.1 **Two climbers scale a sandstone cliff.** *(Photo by E. K. Peters)*

Q8.19. If a geologist tells you that the sand for each 1-foot thickness of this type of rock takes about 300 years to deposit, how much time is represented from the head of the lower climber to the feet of the upper climber? (***Hint:*** You will have to estimate certain things from the photo. Remember that estimates are not just guesses but are serious attempts at performing rough calculations.)

Problem 8.3. Determining Age by Lichen Growth. Primitive organisms called *lichens* (*LIKE-ens*) can aid geologists in assigning absolute dates to recent events. Lichens are a combination of plant and fungus that grow on the surface of rocks. Each lichen begins as a tiny dot and then grows slowly outward, increasing in diameter. The rate at which the diameter grows is complex, neither fully linear nor fully exponential. But the rate is *regular,* which is all we need to assign "lichen dates" to surfaces that have lichens growing on them. Figure 8.2 shows two types of lichens, plus some light areas of rock that have been bleached by chemicals excreted by the lichens.

Q8.20. Using the nickel for scale, find the diameter of the largest single bright yellow lichen. More than one example is acceptable, depending on your reasoning about what constitutes a "separate" lichen. (***Hint:*** Measure a real nickle and correct for the scale of the photograph. Use millimeters for all your measurements.)

Scientists have established growth rates for lichens in various environments. Lichens on building stones in ancient cathedrals, stone bridges, old tombstones, and so on have been used to determine growth curves. A generalized curve is shown in Figure 8.3.

Q8.21. Using this curve, and supposing that the rock in Figure 8.2 was exposed at the surface due to a rockfall or landslide, how long ago did the event occur?

Q8.22. Is this date a minimum or a maximum estimate? Explain.

Problem 8.4. Determining Age by Radioactive Decay. Isotope **A,** with a half-life of 100,000 years, decays to isotope **B.** A magma containing isotope **A** slowly cools and starts to form a mineral that eventually becomes

Figure 8.2   Lichens growing on a granite boulder that has been exposed to the atmosphere for many years. Note nickel for scale. *(Photo by E. K. Peters)*

part of an igneous rock. When the mineral is new (age = 0), it contains only atoms of **A** and no atoms of **B**. But as time passes, atoms of **A** decay within the mineral to form atoms of **B**. (*Important:* Assume that the mineral does not "leak" any atoms to its surrounding environment. All of the **A** and **B** atoms remain a part of the mineral.)

Q8.23. A geologist analyzes the mineral and finds 80 atoms of isotope **A** and 560 atoms of isotope **B.** How old is the mineral?

Problem 8.5. Determining Age by Radioactive Decay. An unstable isotope we will call **Q** has a half-life of 1000 years and decays to a stable isotope we will call **Z.** A

mineral forms from a magma and becomes part of an igneous rock. Initially, it contains no **Z** atoms. A geologist analyzes the mineral and finds a ratio of atoms of **Q** to atoms of **Z** of 1:3.

Q8.24. Assuming no leakage into the environment, how old is the mineral?

Q8.25. If the ratio of **Q** to **Z** were 1:15, how old would the mineral be?

Problem 8.6. Fractional Half-Lives. In the preceding problems and in Lab 8.2, we dealt with whole numbers of half-lives—one, two, three, four, or five full half-lives. However, the more usual case involves a fractional number of half-lives.

The easiest way to solve problems involving fractional half-lives is to use the graph in Figure 8.4 on the following page. This method can be applied to any mineral or rock specimen that contains radioactive elements. The graph's *x* axis shows the number of half-lives that have passed. Its *y* axis shows the present ratio in the specimen of parent isotopes to total isotopes (parent + daughter). The graph assumes that no daughter isotopes were present at the time the mineral formed.

Let's walk through an example using the graph. Imagine that a geologist measures the number of atoms of parent isotopes and the number of atoms of daughter isotopes in a volcanic mineral. She finds 12,000 parent isotope atoms and 8,000 daughter isotope atoms. The ratio (or proportion) of parent atoms to the total (parent + daughter) atoms is

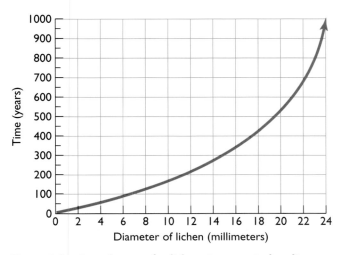

Figure 8.3   Growth curve for lichens in a particular climate.

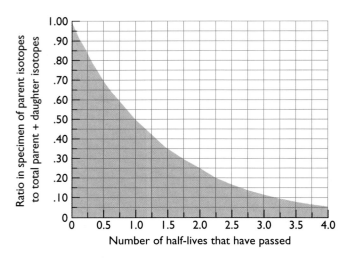

Figure 8.4   Relation between half-lives that have passed and the present ratio in the specimen of parent isotopes to total parent-plus-daughter isotopes.

$$\frac{12{,}000 \text{ parent atoms}}{12{,}000 \text{ parent atoms} + 8{,}000 \text{ daughter atoms}} =$$

$$\frac{12{,}000 \text{ atoms}}{20{,}000 \text{ atoms}} = 0.6$$

Looking at Figure 8.4, trace the 0.60 value on the *y* axis across to the curve drawn on the figure. You can see that approximately 0.74 half-life has passed since the formation of the mineral. Then you can multiply 0.74 by the length of the parent isotope's half-life to find the age of the mineral. (***Note:*** As always, the accuracy of this date rests on the assumption that no parent or daughter isotopes have "leaked" out of or into the mineral since it formed. This assumption has been proven reasonable in many rock samples.)

Now imagine that minerals from different volcanic rocks yield the following data:

Sample	No. of parent isotope atoms	No. of daughter isotope atoms
X	17,010	21,402
Y	971	8,644
Z	8,421	799

Q8.26. If the half-life of the parent isotope is 2.2 million years and if no daughter isotopes were in the minerals when they formed, how old are the minerals in samples **X, Y,** and **Z**?

Problem 8.7. Determining the Age of Minerals in a Rock. Study Figure 8.4 (read problem 8.6 for an intro-

duction if necessary). Then consider a volcanic rock that contains the following:

Mineral grain	No. of thorium-232 atoms	No. of lead-208 atoms
1	1,710	98
2	3,004	187
3	2,101	44
4	806	44

Q8.27. Knowing that the half-life for the decay of thorium-232 to lead-208 is 14 billion years, how old does each mineral indicate the rock to be?

Q8.28. Which mineral shows signs of having lost some daughter isotope atoms at some point?

Problem 8.8. Decay Constant of Uranium-238. Another way of expressing a radioactive element's rate of decay is called the *decay constant,* represented by the Greek letter lambda (λ). The decay constant and the half-life of an element are related to each other by this equation:

$$\text{half-life} = \frac{\ln 2}{\lambda}$$

(***Note:*** "ln 2" denotes the natural log function, or log base *e*. Note also that scientists can measure λ even for long-lived isotopes that have too long a half-life to measure directly.)

Q8.29. The half-life of uranium-238 is 4.47 billion years. What is its decay constant?

Problem 8.9. Equations for Behavior of Radioactive Isotopes. The general equation governing the behavior of radioactive isotope is

$$P_t = P_o e^{-\lambda t}$$

where
P_o = number of parent atoms at starting time
P_t = number of parent atoms at any time
t = span of time
λ = decay constant
e = "the natural number," or the base of natural logarithms, an irrational constant near 2.718

In general, geologists cannot say how many atoms of the parent element initially were present in a mineral or rock. Thus, the above equation is not useful as written. However, we also know that

$$P_o = P_t + D_t$$

where D_t = number of daughter atoms at any time t. (This is simply the equation you worked out for yourself in Lab 8.2.)

Q8.30. On your answer sheet, substitute the foregoing expression for P_o into the formula for P_t.

Q8.31. Using algebra, simplify the equation and solve for t (time). (**Hint:** Your result will have t on the left of the equals sign with no other terms beside it. The equation for time will depend only on variables that geologists can measure in the laboratory.)

Unit 8   Absolute Time

Last (Family) Name_____First Name_____

Instructor's Name_____Section_____Date_____

Q8.1, 8.2, 8.3.

Table 8.1	
Walking times	
Lap (round trip)	**Seconds**
1	
2	
3	

Q8.4.

Figure 8.5   Plot of walking times.

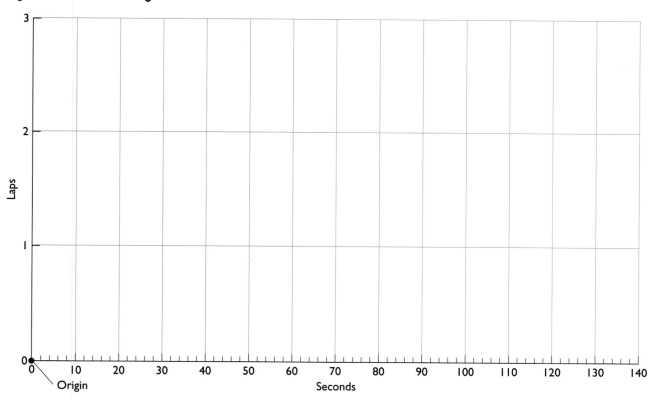

Q8.5. _____ seconds

Q8.6. _____ seconds

Q8.7. _____ gallons

Q8.8. _____ gallons

Q8.9.

Table 8.2

Counting marbles

	Number of marbles in middle box	
	Red	Blue
Start		0
End of Play 1		
End of Play 2		
End of Play 3		

Q8.10, 8.11, 8.12, 8.13.
Figure 8.6 **Plot of marble counts.**

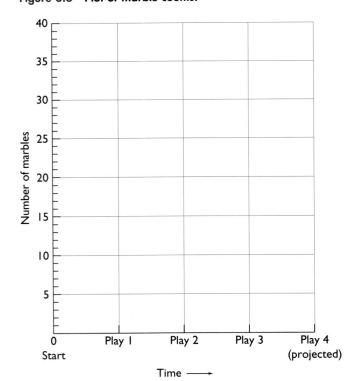

Q8.14. _____ + _____ = _____

Q8.15. _____

Q8.16. _____

Unit 8 Absolute Time

Last (Family) Name_____ First Name_____

Instructor's Name_____ Section_____Date_____

Q8.17. Earth's age:_____ years

Q8.18. The false assumption:_____

_____

Q8.19. _____ years

Q8.20. Diameter (show the unit):_____

Q8.21. _____ years

Q8.22. (circle one) minimum maximum

Explanation:_____

Q8.23. _____ years

Q8.24. _____ years

Q8.25. _____ years

Q8.26.

Mineral in sample **X** is _____ years old

Mineral in sample **Y** is _____ years old

Mineral in sample **Z** is _____ years old

Q8.27.

Mineral **1:** _____ years old

Mineral **2:** _____ years old

Mineral **3:** _____ years old

Mineral **4:** _____ years old

Q8.28. (circle one mineral number) 1 2 3 4

Q8.29. Decay constant: _____

Q8.30. _____

Q8.31. _____

Movers and Shakers

Earth's rocky rind is moving. It may feel solid and motionless, but our planet's crust is never still. It moves at an imperceptibly slow creep. The continents migrate ponderously around the globe, and continent-size portions of the crust move relative to one another. During earthquakes and landslides, the movement becomes a momentary jarring lurch. Ours is most assuredly a dynamic planet.

Plate Tectonics: Geology's Grand Theory — Unit 9

When the idea of "moving continents" first was proposed as a scientific theory in 1912, most geologists laughed. Now, less than a hundred years later, in an amazing about-face, the modern theory of moving continents, called **plate tectonics,** lies at the heart of everything geologists understand about our planet!

Plate tectonics theory emerged in the 1960s, so it still is a young idea. Details of the theory continue to be refined, because science is never done with its work. But the basic theory is well established, and it is worth your most careful study.

Plate tectonics theory explains why Colorado and Tennessee have mountains and why Kansas and Florida are flat. Look around your campus. Plate tectonics theory helps explain why your campus is flat or hilly, why the area has the kind of rocks it does, and even why your region has (or lacks) certain industries. Plate tectonics explains much about our dynamic home, planet Earth.

Lab 9.1   Isostasy

In this experiment, you will explore **isostasy,** a principle that is key to understanding global topography. You will observe and calculate the buoyancy of blocks having different densities. Review isostasy in your textbook for perspective on the variables you will manipulate in this experiment.

Materials
- thick wood block labeled **A**
- thin wood block labeled **B**
- container with flat bottom and clear sides, half-filled with water
- ruler
- paper towels
- mass balance
- optional: calculator

Procedure
1. Weigh each block carefully, in grams.

Q9.1. Record your results in Table 9.2 on your Lab Answer Sheet.

2. Calculate the volume of each block. To do so, measure the length, width, and height of each block in centimeters. (**Note:** The blocks may not be perfectly square, but make your measurements as accurately as you can.) Multiply length × width × height to calculate the volume for each block.

Q9.2. Record your answers in Table 9.2.

3. Calculate the density (weight per unit volume) of each block.

Q9.3. Record the density in Table 9.2.

4. Float block **A** in the water. (It should float evenly and not be submerged more on one side than on

another. If you are not sure your block is floating in a level manner, consult your instructor.) Precisely measure the vertical dimension of the block that extends *below* the water line. (For example, this amount might be several millimeters.)

Q9.4. Record your result to the nearest half millimeter (0.5 mm) in Table 9.3 on your Lab Answer Sheet.

5. Do the same for the vertical dimension that extends *above* the water line.

Q9.5. Record your results as indicated in Table 9.3.

6. Now remove block **A** from the water. Dry it completely to avoid water absorption into the block.

7. Calculate the percentage of the vertical height of the block that is above the water line. Do the same for the percentage below the water line. The two percentages should add to 100% (or very close).

Q9.6. Record your percentage calculations as indicated in Table 9.3.

8. Repeat Steps 4–7 for block **B**.

9. Now, look for regularities in your results. Remember, the density of water is 1.0 gram per cubic centimeter.

Q9.7. Using that fact and your earlier results, complete Table 9.4 on your Lab Answer Sheet.

10. Are the last two columns of Table 9.4 in good agreement? If not, consult your instructor. The densities of the blocks should clearly reflect how they floated in the water.

11. Now we'll move to another important isostasy concept. Float block **A** in the water. When it has stabilized, gently place block **B** on top of block **A**.

Q9.8. Roughly what proportion of block **A** is now below the waterline?

12. Gently lift block **B** off block **A**.

Q9.9. Very approximately, how long did it take, in seconds, for block **A** to adjust (achieve isostasy) once you removed block **B**?

13. Completely dry both blocks.

Applying What You Just Experienced

As you have learned in lectures, blocks of Earth's crust behave as if they were "floating" on much deeper Earth materials. The principle of isostasy helps explain how blocks of crust that have different densities "float" either higher or lower. Continental crust is both less dense and thicker than oceanic crust.

Q9.10. Which of the blocks you worked with is the better analogy for continental crust?

Q9.11. Recalling that oceanic crust is the most dense crust on the Earth, which of the blocks you worked with is the better analogy for oceanic crust?

Q9.12. During the Pleistocene Era (roughly equal to the Quaternary Period of geologic time, and commonly known as the "Ice Age"), an ice sheet moved across the continental crust in what is now Scandinavia (Norway, Sweden, and Finland). The massive ice sheet actually depressed the continental crust. What did you do with the wood blocks that was analogous to this event?

Q9.13. How did the rate of our wood block experiment differ from the rate of the real geologic process of glaciation?

Lab 9.2 Pressure and Density Changes within Earth

Minerals in Earth's mantle experience much greater pressure than rocks in the continental crust. This tremendous difference in pressure is evident in the density of two well-known forms of carbon: graphite and diamond.

Materials
- piece of graphite
- graduated cylinder
- water
- mass balance

Procedure

1. Weigh your piece of graphite, handling it carefully so that it does not break apart.

Q9.14. Record the weight in Table 9.5 on your Lab Answer Sheet.

2. Add some water to your graduated cylinder. Note the volume here: _____. Tip the cylinder to about a 45° angle and gently slide the graphite into the water so that no water splashes out of the cylinder. (If any does, start over.)

3. Air bubbles may cling to the graphite. Tap the graduated cylinder to dislodge the air bubbles so that they rise to the surface and rejoin the atmosphere.

4. Note the volume of the water plus the graphite in the graduated cylinder: _____. Subtract the original volume of water you measured.

Q9.15. Record the result in Table 9.5. This is the volume of the graphite. (Remember that 1 milliliter = 1 cubic centimeter.)

5. Calculate the density of your piece of graphite (density = weight/volume).

Q9.16. Record your result in Table 9.5.

Applying What You Just Experienced
If your instructor had a huge budget, you would have been given a large diamond and asked to find its density. But we live in the real world, so we'll just tell you the density of diamond: 3.5 grams per cubic centimeter.

Q9.17. How many times greater than the density of your graphite sample is the density of a diamond?

Q9.18. Diamond is virtually pure carbon. Graphite is a bit less pure, but for our purposes, we will treat it as pure carbon. Which form of carbon, diamond or graphite, packs its carbon atoms closer together?

Q9.19. Diamond, which comes from approximately 200 kilometers below our planet's surface, is much denser than graphite, which forms near the surface in continental crust. How does Le Châtelier's principle explain this? (See Unit 1 for a discussion of Le Châtelier's principle.)

Q9.20. Imagine placing your piece of graphite in a device that can generate conditions equivalent to those at a depth of 200 kilometers into Earth's mantle—the region where diamonds can form. This device converts your piece of graphite into diamond. What would be the resulting volume of your new gemstone? (**Hint:** Consider the density of diamond given above.)

Q9.21. Knowing that 1 gram = 5 carats, how many carats of diamond would your graphite produce?

Lab 9.3 Relative and Absolute Motion

Earth's lithospheric plates move slowly around the planet. They move at different rates, usually only centimeters per year. If we measure movement between two moving plates, we are measuring *relative* plate motion. If we measure a plate's movement compared to a fixed point, we are measuring *absolute* motion.

Materials
- student A and student B
- tape measure, yardstick, or ruler
- plastic overlay sheet
- marking pen

Procedure
1. As a group, determine which direction in your lab room is north.

2. Student A should stand 10 feet to the north of student B.

3. Student A should now step 3 feet southward and stop.

4. Student B next steps 1 foot northward and stops.

Q9.22. How many feet are now between student A and student B?

Q9.23. The change in distance between student A and student B is an example of what type of motion?

5. Notice that the net distance change between students A and B occurred faster than either of their individual movements.

Q9.24. Imagine that student A was standing on a lithospheric plate and student B was standing on a different plate. Somewhere between them there must be a plate margin (a contact line between two plates). In the analogy just enacted, what type of margin would it be?

6. Notice that your answers to the preceding questions would not change even if your entire lab room were slowly creeping in any direction. (Indeed, your lab room *is* moving, because it is riding on one of Earth's lithospheric plates.)

7. Now let us consider lithospheric motion in *absolute* terms, relative to a fixed point. Imagine a *hot spot*, or source of magma, deep within Earth beneath the lithospheric plates. Assume that the hot spot is fixed in place in Earth's mantle.

8. As an analogy for the hot spot, student A should hold a marking pen pointing straight up—and absolutely still.

9. Student B next holds the plastic overlay horizontal and above the marking pen, just touching the pen.

10. Student B then slowly moves the overlay north.

Q9.25. This activity draws a line on the overlay. The line grows longer (is drawn) toward which compass direction?

11. The plastic overlay is an analogy for an oceanic plate. The line made by the marking pen is an analogy for a chain of volcanoes.

Q9.26. If an oceanic plate moves west over a hot spot, in which compass direction will the chain of volcanoes over the hot spot grow over time?

Applying What You Just Experienced

Californians living between Los Angeles and San Francisco know a good bit about relative motion, courtesy of the San Andreas Fault. The fault is a strike-slip or transform boundary that runs northwest–southeast. Land west of the fault is moving northward relative to land east of the fault. The land does not move all the time, however. It tends to lurch during earthquakes and move hardly at all between quakes. The biggest movement along the San Andreas Fault in recorded history occurred near San Francisco in 1906.

Q9.27. Imagine that student A and student B stood facing each other across the fault in 1906 when the earthquake struck. Each student would

have seen the other move about 20 feet during the earthquake. Student A would say, "B moved 20 feet to my right," and student B would say, "A moved 20 feet to my _____."

Note that the *relative* motions of students A and B are clear, but that they cannot report *absolute* motion based on what they saw during the quake.

Now consider the motion of the Hawaiian Islands. They are a chain of volcanoes that have formed and continue to form as the Pacific Plate moves very slowly over a hot spot deep in Earth's mantle. For our purposes, we will assume that the hot spot is stationary.

The Hawaiian Islands include the islands of Hawaii, Maui, Molokai, and others. The youngest island in the Hawaiian chain is the "Big Island," Hawaii (Figure 9.1). Maui is older than Hawaii, Molokai is older than Maui, and so on looking northwestward.

Draw a straight line on Figure 9.1 from the center of the "Big Island" through the center of Oahu to the edge of the figure. The line you have drawn approximates the motion of the oceanic plate over the hot spot that is generating the islands.

Q9.28. The **volcanism** grows younger (more recent) in which general compass direction?

Q9.29. Recalling your work with the plastic overlay and marking pen, in which compass direction has the Pacific Plate been moving?

Q9.30. Is Maui moving relative to Molokai?

Q9.31. Are both Maui and Molokai moving in absolute terms?

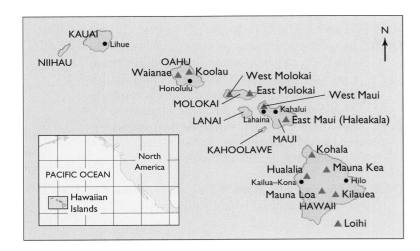

Figure 9.1 Hawaiian Island chain and major volcanoes. *(Courtesy of U.S. Geological Survey)*

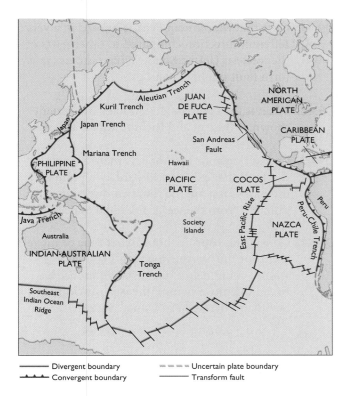

--- Divergent boundary   - - - Uncertain plate boundary
▲▲▲ Convergent boundary   —— Transform fault

Figure 9.2   Earth's lithospheric plates. *(After F. Press and R. Siever,* Understanding Earth, *2d ed. New York: W. H. Freeman and Company, 1998, front endpaper)*

Problems

Problem 9.1. Testing Plate Tectonics Theory Worldwide.
Plate tectonics theory has been tested in several ways. Changes in distance between points on Earth's surface have been measured by satellite (Table 9.1).

Figure 9.2 is a map showing the Pacific Ocean. Study it and the table together. Consider the sense of change (positive for farther apart, negative for closer together) for the four pairs of points in the table.

Q9.32. In the first two examples, Hawaii to Australia and Hawaii to Japan, do the location and types of plate margins explain the sense of change (positive or negative) in distance?

Q9.33. Considering a more complex situation, name the types of plate boundaries between Hawaii and Peru.

Q9.34. The sense of change between Hawaii and Peru indicates that which plate margin is overwhelming the effects of the other?

Q9.35. Look at the information you have about Hawaii and the Society Islands. Would the figure lead you to predict any change in distance between the two? Explain.

Q9.36. Is the magnitude (size) of the change in distance between these two island groups large or small compared to the other information in the table?

Q9.37. What might account for the data concerning the change in distance between the two island groups? (More than one answer is possible.)

Table 9.1

Satellite data of distance changes between points on Earth's surface

Points on Earth's surface	Change in distance	Sense of change
Hawaii to Australia	9.0 ± 0.6 centimeters per year	Closer together (negative)
Hawaii to Japan	6.1 ± 0.1 centimeters per year	Closer together (negative)
Hawaii to Peru	7.9 ± 0.3 centimeters per year	Farther apart (positive)
Hawaii to Society Islands	0.9 ± 0.6 centimeters per year	Farther apart (positive)

Problem 9.2. Testing Plate Tectonics Theory in the United States. Figure 9.3 shows the contemporary movement of certain spots within the United States, as measured by sophisticated satellite techniques.

- The *magnitude* (size) of movement is shown by the *length* of the arrows on the figure.
- The *direction* of movement is shown by the *direction* of the arrow.

Q9.38. From the data given, what place shows the greatest movement?

Q9.39. How many times greater is this movement compared to that experienced at Richmond, Florida?

Q9.40. What is the compass direction of the movement shown at Bear Lake, Platteville, McDonald Observatory, Owens Valley, Quincy, and Monument Peak?

Q9.41. Note that Monument Peak is headed in a very different direction compared to other parts of the western United States. What famous transform fault might explain this?

Q9.42. What is the compass direction of movement shown at Boston and Richmond?

Q9.43. Consulting your textbook, is there a single, *active* fault that explains why the two points on the East Coast are moving in a direction quite different from the western United States?

Q9.44. Consulting your textbook, what type of movement for the North American Plate (as a whole) could help explain the difference between the movement in the eastern United States and the western United States?

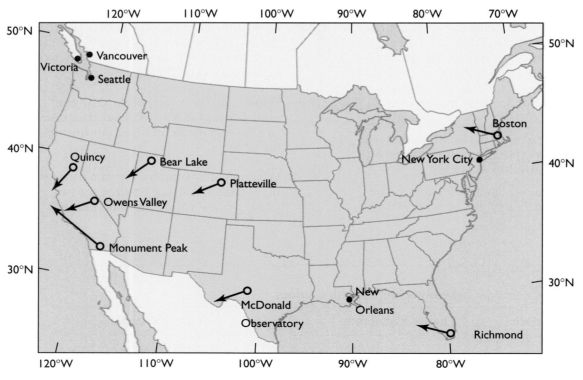

Figure 9.3 Tectonic motion within the United States. *(SLR data, modified from http://cddisa.gsfc.nasa.gov/926/slrtecto.html)*

Problem 9.3. Volcanic Features in the Lower 48 States. Figure 9.4 shows volcanic eruptions in the 48 contiguous states during the past 2000 years. Also shown are other, potentially active volcanic areas.

Q9.45. One area with significant volcanic activity in the past 2000 years is in Oregon and Washington, as well as northernmost California. Referring to your textbook or a map, what is the name of the volcanic mountain chain in these states?

Q9.46. What type of plate margin is associated with these volcanoes?

Q9.47. The region from Los Angeles to San Francisco, along the coast of California, is cut by a famous transform (strike-slip) fault. What is the name of this fault?

Q9.48. Is this fault clearly associated with volcanic activity in the past 2000 years?

Q9.49. Does the eastern United States show any volcanic activity in the past 2000 years?

Q9.50. Is a *plate* margin represented by the eastern United States? (Refer to your textbook.) What do geologists call such a continental margin?

Q9.51. Is it likely that Boston, New York, and Atlanta will be directly affected by volcanic eruptions?

Q9.52. Could people in Seattle experience some direct effect from volcanic eruptions?

Q9.53. Movies sometimes portray Los Angeles as being in jeopardy from a volcanic eruption. Study the map and decide for yourself: Is Los Angeles within 100 miles of any volcanic feature shown on the map?

Problem 9.4. Some Ages of Oceanic Rocks. Early in the twentieth century, geologists thought of the continental crust as stable and fixed. They knew that some rocks on the continents were ancient, while others were relatively young. Since no obvious rock-forming processes occurred on the deep ocean floor, they assumed that the age of the rocks in the oceanic crust must be enormous—likely from the Archean Eon or Proterozoic Eon.

In the 1950s, the technology for drilling into oceanic rocks from ships improved markedly. For the first time, geologists acquired samples of rocks from beneath the deep ocean. Figure 9.5 on the following page shows the age of seafloor rocks worldwide as we know it today.

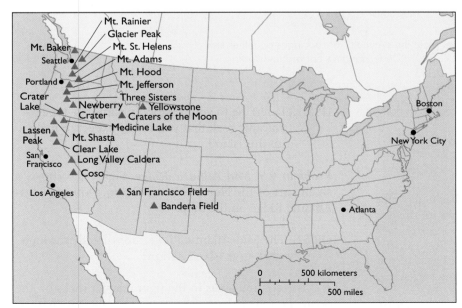

▲ Volcano active during past 2000 years   ▲ Other potentially active volcanic areas

Figure 9.4   Potentially active volcanoes of the western continental United States.
(After L. Topinka, U.S. Geological Survey/CVO, 1999; from S. R. Brantley, 1994, Volcanoes of the United States, *USGS General Interest Publication.)*

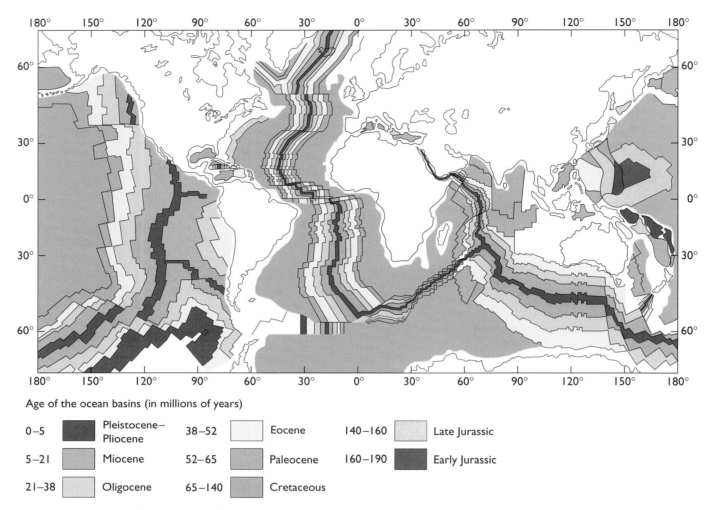

Age of the ocean basins (in millions of years)

0–5	Pleistocene–Pliocene	38–52	Eocene	140–160	Late Jurassic
5–21	Miocene	52–65	Paleocene	160–190	Early Jurassic
21–38	Oligocene	65–140	Cretaceous		

Figure 9.5   Age of seafloor rocks worldwide. Note the symmetry of rock ages on either side of the Mid-Atlantic Ridge. (*From F. Press and R. Siever,* Understanding Earth. *New York: W. H. Freeman and Company, 1994, Figure 20.11)*

Q9.54. From what *eon* are the oldest oceanic rocks?

Q9.55. From what *era* are these oldest rocks?

Q9.56. From what *period* are these oldest rocks?

Q9.57. Looking at the Atlantic Ocean basin, describe the location of the youngest rocks on the ocean floor.

Q9.58. Where are the increasingly older rocks of the Atlantic, compared to these young rocks?

Q9.59. What is the geologic period and age range of the oldest rocks in the Atlantic basin?

Q9.60. What well-known island in the North Atlantic contains the youngest rocks of the ocean crust?

Q9.61. Referring to your textbook, name the type of volcanic rock that you would expect to make up this island.

Problem 9.5. Mid-Atlantic Ridge. Figure 9.6 shows a small part of the North Atlantic Ocean. Note the Mid-Atlantic Ridge on the map.

Q9.62. The Mid-Atlantic Ridge runs directly through what large island nation?

The colored areas in the figure depict rocks on the ocean floor that have a "normal" magnetic signature, so-called because it matches the current direction of polar magnetism. Between the colored lines are narrow blanks. These indicate rocks that have a "reverse" direction of magnetism from that existing today.

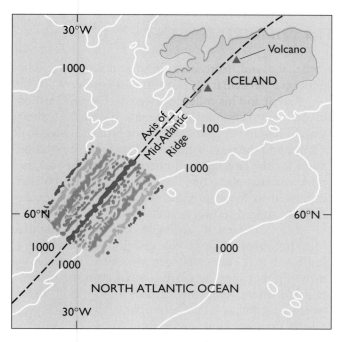

Figure 9.6   Mid-Atlantic Ridge. *(From F. Press and R. Siever, Understanding Earth, 2d ed. New York: W. H. Freeman and Company, 1998, Figure 20.9)*

Q9.63. A single color (such as green) indicates rocks of the same age. Referring to the figure, does it appear that the spreading of the Mid-Atlantic

Ridge has generated a fairly equal amount of oceanic crust on either side of the ridge?

When Vikings settled Iceland late in the tenth century, or 900's, they gathered once a year at a particular place on the island to discuss problems and to make decisions. Icelandic society thus became the first democratic republic in European history. The meeting place for these annual discussions is directly on the axis of the Mid-Atlantic Ridge. It was selected because of a high natural "speaker's platform" that rose on the western side of the ridge. Below and to the east, across the narrow rift valley, listening Vikings could hear what the speaker had to say (Figure 9.7).

Q9.64. Referring to your textbook, what is the name of the tectonic plate on which the Viking speaker stood?

Q9.65. What is the name of the tectonic plate on which the listeners were gathered?

Q9.66. If modern Icelanders reenacted a Viking meeting today, with the speaker and audience standing on the same rocks where their ancestors stood, would the speaker be the same distance from the audience, closer to them, or farther from them? Why?

Figure 9.7   Viking meeting place in Iceland, named Thingvellir (the Thing Place). We are looking right down the axis of the Mid-Atlantic Ridge, the low area where two vast tectonic plates are parting company. Rocks on the left are moving westward, and rocks on the right are moving eastward. *(Photo by Pamela Judson Rhodes)*

Geo Detectives at Work   THE CASE OF THE MISBURIED AGENT

In 1985, an agent of the U.S. Drug Enforcement Agency (DEA) was kidnapped, tortured, and murdered in Guadalajara, Mexico. His body was not immediately found, so the U.S. government put diplomatic pressure on the Mexican Federal Judicial Police (MFJP) to investigate the murder and arrest the drug runners involved. Several weeks went by with no results. Then, suddenly, the MFJP produced the agent's body. They said it was discovered on a ranch in Mexico's mountains. The ranch belonged to a small-time drug-smuggling family with the last name of Bravo. The MFJP said that their raid on the Bravo ranch had killed every member of the family. And the family must have killed the American agent. Case closed.

Or was it? An American geologist named Ron Rawalt worked for the FBI. He didn't buy the MFJP's story.

Rawalt knew that Mexico's mountains, produced by tectonic forces and volcanic processes, have many different soils. With Rawalt in charge, an FBI investigation quickly established that soil particles from the Bravo ranch contain obsidian grains, dark and rounded. But the soil scraped from the DEA agent's body was rich in light-colored tuff and ash. This pointed to burial somewhere else in the mountainous tectonic range of central Mexico.

Closer examination showed that the light-colored material contained two kinds of cristobalite, an unstable form of quartz. Cristobalite occurs in silica-rich volcanic lavas, generally of Cenozoic age. Unfortunately, Mexico's mountains have tremendous volumes of silica-rich volcanics, produced in response to the tectonic forces throughout the Cenozoic Era. Clearly, the agent's body had been buried elsewhere than at the Bravo ranch, but where? This was a needle-in-a-haystack situation.

Luckily, one more piece of geologic evidence appeared in the light-colored soil. It was a few grains of volcanic glass tinged an unusual deep pink color. A geologist at the Smithsonian Institution in Washington recognized these unusual volcanic particles. They

come, she said, from an area in and around a state park on a particular Mexican mountainside.

Rawalt and his colleagues went to the park, armed in case any corrupt MFJP agents objected to their search. Rawalt set up his microscope on the tailgate of a truck and the team brought soil samples. He studied them for the two kinds of cristobalite grains and the pink volcanic glass. But even though the search had narrowed to the state park, there were many square miles of rugged terrain to search!

Then a French geologist noted the slightly rounded shape of some volcanic grains. Their shape indicated that they had been deposited by water—meaning that they came from a place on the mountainside where water flowed, like a gully. Still, there were many gullies on the mountainside. At this point, a local man who knew something about the murder decided to help. He directed the FBI team to one particular gully. The soil in the gully featured all three mineralogical clues found on the DEA agent's body.

Sniffer dogs quickly found the exact spot in the gully where the agent had been buried before he was unearthed and dumped at the Bravo ranch. Clearly, some members of the MFJP had been involved in the agent's murder. When diplomatic pressure on the MFJP became intense, however, they had unearthed the corpse from its hiding place, "planted" it on the Bravo ranch, and killed everyone there in order to pin the crime on someone else. But forensic geology had shown that the Bravo drug runners were innocent of the DEA agent's murder.

Q9.67. In Mexico, tectonic forces have raised a chain of mountains made of what rock type?

Q9.68. In the state park, the DEA agent's body was temporarily buried in soil that contained _____, an unusual form of the mineral_____.

Q9.69. Would the mineral you named for the preceding question be present in mafic lava flows?

Unit 9   Plate Tectonics

Last (Family) Name_____First Name_____

Instructor's Name_____Section_____Date_____

Q9.1, 9.2, 9.3.

Table 9.2			
Weights and densities of blocks			
Block	**Weight in grams (g)**	**Volume in cubic centimeters (cm^3)**	**Density in grams per cubic centimeter (g/cm^3)**
A			
B			

Q9.4, 9.5, 9.6.

Table 9.3		
Flotation results		
	Block A	**Block B**
Above waterline in millimeters (mm)		
Below waterline in millimeters (mm)		
Total above + below		
Above waterline (percent)		
Below waterline (percent)		

Q9.7.

Table 9.4				
Density calculations				
Block	**Original density of the block in grams per cubic centimeter (g/cm^3)**	**Density of water minus the density of block in grams per cubic centimeter (g/cm^3)**	**Convert decimal to percent**	**Proportion of block you measured above waterline (percent)**
A				
B				

Q9.8. _____ %

Q9.9. _____ seconds

Q9.10. _____

Q9.11. _____

Q9.12. _____

Q9.13. _____

Q9.14, 9.15, 9.16.

Table 9.5

Weight, volume, and density of graphite sample

Weight in grams (g)	Volume in cubic centimeters (cm³)	Density in grams per cubic centimeter (g/cm³)

Q9.17. Diamond's density is _____ times greater than that of my graphite sample.

Q9.18. (circle one)   diamond   graphite   can't tell

Q9.19. _____

_____

Q9.20. _____ cm³

Q9.21. _____ carats

Q9.22. _____ feet

Q9.23. (circle one)   relative motion   absolute motion   can't tell

Q9.24. (circle one)   convergent   divergent   transform

Q9.25. (circle one)   N   S   E   W

Q9.26. (circle one)   N   S   E   W

Q9.27. (circle one)   left   right   neither   both

Q9.28. (circle one)   NE   SE   NW   SW

Q9.29. (circle one)   NE   SE   NW   SW

Q9.30. (circle one)   yes   no

Q9.31. (circle one)   yes   no

Unit 9   Plate Tectonics

Last (Family) Name_____First Name_____

Instructor's Name_____Section_____Date_____

Q9.32. Hawaii to Australia: (circle one) yes_____ no _____

Hawaii to Japan: (circle one) yes_____ no _____

Q9.33. _____ and _____

Q9.34. _____

Q9.35. _____

_____

Q9.36. (circle one)   large   small

Q9.37. _____

Q9.38. _____

Q9.39. _____ times greater

Q9.40.

Bear Lake: _____

Platteville: _____

McDonald Observatory: _____

Owens Valley: _____

Quincy: _____

Monument Peak: _____

Q9.41. _____

Q9.42.

Boston: _____

Richmond: _____

Q9.43. (circle one)   yes   no

Q9.44. _____

Q9.45. _____

Q9.46. _____

Q9.47. _____

Q9.48. (circle one)   yes   no

Q9.49. (circle one)   yes   no

Q9.50. (circle one)   yes   no

Geologists call this a _____

Q9.51. (circle one)   yes   no

Q9.52. (circle one)   yes   no

Q9.53. (circle one)   yes   no

Q9.54. _____ Eon

Q9.55. _____ Era

Q9.56. _____ Period

Q9.57. _____

Q9.58. _____

Q9.59. _____ Period, _____ to _____ million years old

Q9.60. _____

Q9.61. _____

Q9.62. _____

Q9.63. (circle one)   yes   no

Q9.64. _____

Q9.65. _____

Q9.66. (circle one)   the same   closer   farther

Why? _____

Q9.67. (circle one)   volcanic   sedimentary   metamorphic

Q9.68. _____ , _____

Q9.69. (circle one)   yes   no   can't tell

Earthquakes are a major natural hazard in many countries. They are also of special interest to scientists because many are caused by the movement of Earth's great tectonic plates. The worldwide pattern of earthquakes provides strong supporting evidence for plate tectonics theory.

During your lifetime, you or a family member will likely feel seismic waves moving through the rock and soil beneath your feet. The experience can be bewildering, thrilling, or terrifying, depending on the intensity and duration of the quake. Major earthquakes can be tragic, with hundreds or thousands of people buried in the rubble of collapsing buildings or burned in the fires that commonly follow a major quake. As we have seen in recent years in California, property damage from a single earthquake can easily run into the billions of dollars.

In this lab, you will work with miniature analogs of earthquake waves and use a simple technique to locate the epicenter of earthquakes.

Lab 10.1 Measuring Velocities of Different Types of Waves

In this experiment, you will manipulate analogs (simulations) for different types of seismic waves. You will observe how they travel and how their different speeds let us determine where an earthquake originated.

Materials
- tightly coiled long spring (fairly rigid)
- more loosely coiled spring (like a Slinky™)
- stopwatch
- tape measure or yardstick/meterstick

Procedure
1. Your instructor will form teams of three to five people, assigning two *springholders*, a *timekeeper* (someone with quick reflexes and a knack for timing), a *measurer*, and a *data recorder*. (The last three functions can be combined.)

2. Move to where you will have ample space to perform this experiment, or to wherever your instructor directs you.

CAUTION Springs are easily damaged. Do not overextend or otherwise deform them.

3. *Springholders:* Start with the looser spring. Hold opposite ends of it, keeping the entire length of the spring off the floor. Move apart, extending the spring to about double its length (or as your instructor directs). Then stand still, in the same spot, throughout the experiment. See Figure 10.1 on the following page.

4. *Measurer:* Measure the distance on the floor between the two ends of the spring.

Q10.1. On your Lab Answer Sheet, record the length of the extended spring in Table 10.1. (***Note:*** Measure on the floor from one end of the spring to the other. Don't follow the drooping arc of the spring with the tape measure.)

5. *Timekeeper:* Stand beside one of the *springholders. Springholder nearest the timekeeper:* Grasp two or three coils of the spring and compress them together so they touch. When the *timekeeper* gives the signal, abruptly release the bunched coils (but hang onto the end of the coil!).

6. *Timekeeper:* Start timing (in seconds) the instant the bunched coils are released. Closely observe the wave that travels through the spring. The wave travels to the far end of the coil, reflects against the hands of the *other springholder,* and then returns to the *first springholder.* At this moment, stop timing. (In other words, measure the number of seconds for one round-trip of the wave from one end of the spring to the other and back.) The wave you are creating and measuring simulates a seismic **P-wave,** or **compression wave.**

7. *Team:* Repeat this procedure at least three times until everyone gets the hang of it and you obtain repeatable results. Tell your instructor when you think you are achieving consistent timing.

Q10.2. When your instructor approves, the *data recorder* records the round-trip compression wave time in Table 10.1.

8. *Springholder nearest the timekeeper:* This step requires some practice. Holding one end of the spring, *sharply and quickly* jerk your hand up and down, one time, enough to move the end of the spring up and

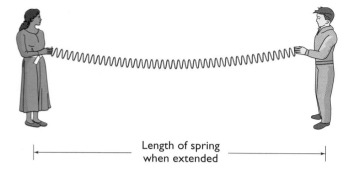

Length of spring
when extended

Figure 10.1 Springholders ready for seismic wave experiment.

down 3 or 4 inches. The goal is to initiate a single **"S-wave"** or "shear wave" that travels down the spring and back. (**Note:** Don't make a big wave—it will result in sloppy time measurements.) Practice until you can make a single, sharp, relatively small wave. Now make the wave when the *timekeeper* tells you to.

9. *Timekeeper:* Time the interval (in seconds) required for the wave to travel from one person to the other and back.

Q10.3. When your team achieves repeatable results, get your instructor's approval to record your round-trip shear wave time in Table 10.1.

10. *Team:* Using the smaller, tighter spring, repeat steps 3–9.

Q10.4. Record your team's results in Table 10.1.

11. Share your data so that everyone on your lab team has a fully completed Table 10.1.

Applying What You Just Experienced

For each spring, the first wave type you made is like a seismic **P-wave** (primary wave, with compressional motion). The second wave type you made is like a seismic **S-wave** (shear wave, or secondary wave, with up-and-down motion). Real rocks, of course, do not behave quite as simply as the spring coils you just observed. But the experiment gives you a general idea of how different types of seismic waves move through Earth.

Look at your data. You can see that compression waves in springs travel at a different rate than shear waves. The same is true in Earth's rocks. When a real

earthquake occurs, faster-moving P-waves travel from the epicenter (source of the quake), outpacing the S-waves. Thus, the P-waves arrive first at seismic monitoring stations, providing the first notice that a quake has occurred. The S-waves arrive on the heels of the P-waves. Why is this important? Because the *difference in arrival time* of P-waves and S-waves at seismic stations allows you to calculate the distance to the source of the seismic waves (the epicenter).

Q10.5. In your experiment with the looser spring, what was the speed of your "P-wave"? (**Hint:** Remember to double your length measurement, because your time is a round-trip measurement.)

Q10.6. What was the speed of your "S-wave?"

Q10.7. Which was faster, your "P-wave" or your "S-wave"?

Q10.8. Let's say that the library is about 800 meters from your lab room. How many seconds would it take for your "P-wave" to travel from the lab to the library?

Q10.9. How long would it take for your "S-wave" to travel from your lab to the library?

Q10.10. How many seconds difference is there between the time for your "P-wave" to arrive at the library and the time for the "S-wave" to arrive there?

Geologists and engineers use the difference in the speeds of P-waves and S-waves to generate short-term warnings of impending seismic hazard. This works for two reasons: in Earth rocks, P-waves always arrive sooner than S-waves, and P-waves do relatively little damage compared to S-waves. This allows a very brief interval to prepare for the arrival of the more destructive S-waves.

For example, in earthquake-prone Japan, electronic P-wave monitors are stationed at regular intervals along railroad tracks used by high-speed trains. When a P-wave arrives at any station, all power to the trains is cut immediately. The hope is that the trains will come to a stop before the S-waves arrive and destroy structures like tracks and bridges. In the earthquake that struck Kobe, Japan, in the early 1990s, automatic P-wave detectors were very effective.

Seismic waves travel quickly, but their speed is not entirely constant. Slight *velocity differences* can be detected when seismic waves travel through different types of rock. (Velocity refers to the rate and direction of motion.)

Q10.11. In your experiment, did you observe velocity differences for the "P-waves" in the two types of springs? Explain what you observed.

Lab 10.2 Liquefaction of Wet Sediment During Earthquakes

Unconsolidated (loose) sediment is unstable during earthquakes. Even worse is *wet* unconsolidated sediment. This exercise presents you with an interesting analogy for the liquefaction that may occur in loose sediment during quakes.

Materials
- cornstarch
- water
- mixing bowl
- mixing spoon
- measuring cup (any cup, like a coffee cup)
- two ball bearings or marbles

Procedure
1. Place a cup of cornstarch in the bowl. Don't pack the cornstarch in the cup—just fill the cup to the rim and level it off.

2. Slowly add about one-third cup of water while stirring the cornstarch. Your goal is to make a smooth, mudlike substance, with no soupy parts and no stonelike lumps. (**Note:** Parts of the cornstarch will turn into a hard, stonelike substance first. Keep stirring, chipping away at the hard material, and scraping loose cornstarch off the sides of the bowl.) If necessary, add a few more drops of water. When the consistency seems right, get your instructor's approval to proceed.

3. Using your bare hands, form the material into a ball about the size of a tennis ball. Keep your hands moving, shifting the sphere in your palms and keeping a light pressure on it to hold the material together. When the ball is well formed, rest it on a tabletop. Watch the ball closely.

Q10.12. Describe what you see on the Lab Answer Sheet.

4. Divide the ball into roughly equal portions—one piece for each person in your group. Shape your piece into a ball. Then quickly tear the ball in half and study the ruptured surface.

Q10.13. What does the broken surface of the ball look like the instant you tear it?

Q10.14. What does the same surface look like a second or so later?

5. Return all pieces of the mixture to the bowl and form them into a single mass. You may need to add a little water until you restore a smooth, mudlike consistency throughout the material.

6. Shape the mixture roughly into a ball again and set it on a tabletop. Gently press the material on the tabletop, making a patty about a half-inch thick.

7. Roll a ball bearing or marble across the cornstarch patty. Keep pushing—the cornstarch is sticky, so the marble or bearing will not roll freely. Notice that you can keep the marble or bearing on the surface of the cornstarch as long as it is moving.

8. Now gently rest the ball bearing or marble on the cornstarch patty. Watch it closely for 30 seconds.

Q10.15. Describe what happens to the bearing or marble that is sitting still.

Applying What You Just Experienced
The behavior of the wet cornstarch mixture is roughly like the behavior of a soil-sediment mixture during an earthquake. When seismic waves shake Earth materials, the materials can become liquefied and behave like the wet cornstarch did without being shaken. This liquefaction changes the properties of solid ground dramatically, from those of a solid to those of a thick or viscous liquid. Your textbook probably discusses this phenomenon in the section on earthquake damage. Note that a strong person may be able to run on top of liquefied material without sinking, but buildings and weaker people sink into the sediment or soil.

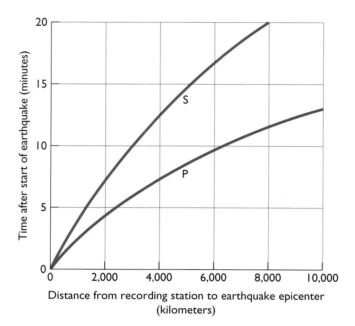

Figure 10.2 Travel-time curves for P- and S- waves.
Earthquake P-waves and S-waves are plotted over time and distance. The different rates of travel are the key to pinpointing a quake's epicenter.

Problems

Problem 10.1. Translating P- and S-Wave Gaps into Distance. In Table 10.2 on your Problem Answer Sheet, you see four time gaps between P-waves and S-waves. For each of these, use Figure 10.2 to calculate the distance from the recording station to the earthquake's epicenter. (***Hint:*** Measure the gaps *vertically* between the P-wave and S-wave curves. For example, in the figure, measure along the vertical lines.)

Q10.16. Record the distances in Table 10.2 on the Problem Answer Sheet.

Problem 10.2. Translating a Distance into a P-Wave/S-Wave Gap. Refer to Figure 10.2. This time, we'll use it backward.

Q10.17. Suppose an earthquake occurs 1000 kilometers from a seismograph. What difference in P-wave and S-wave arrival time would you expect?

Q10.18. Suppose an earthquake occurs 2000 kilometers from a seismograph. What difference in P-wave and S-wave arrival time would you expect?

Problem 10.3. Planet Xenon Has Earthquakes, Too. Imagine that you are the science officer aboard a space-craft. You visit a planet far beyond our solar system, and you name it Xenon. You discover that Xenon has rocks quite different in composition and density from Earth's. After conducting a seismic test, you realize that S-waves on Xenon travel exactly as fast as P-waves.

Q10.19. Could you locate the epicenter of an earthquake on Xenon using the same method you've just been taught? Why or why not?

Q10.20. Knowing the speed of P-waves and S-waves on Xenon, how could you determine the epicenter of Xenon quakes?

Problem 10.4. Seismic Risk in the United States. The U.S. Geological Survey map in Figure 10.3 indicates approximate seismic risk in the United States. This map is based on probability rather than certainty.

Q10.21. Assign the numbers 1 through 7 to the hazard scale in Fig. 10.3, with 1 being the lowest hazard. What level of hazard is expected for future earthquakes at or near Seattle, Los Angeles, Chicago, Miami, and New York?

Q10.22. What future earthquake risk may be expected for your college or university?

Q10.23. What is your hometown (city and state)? What earthquake risk is expected for your hometown?

Problem 10.5. Magnitudes of Different Earthquakes. Review the Richter scale, used to measure the magnitude of earthquakes.

Q10.24. Imagine that Los Angeles experiences an earthquake of Richter magnitude 4.5 today. Tomorrow, San Francisco experiences an earthquake of Richter magnitude 6.5. How many times larger in ground motion is the San Francisco earthquake?

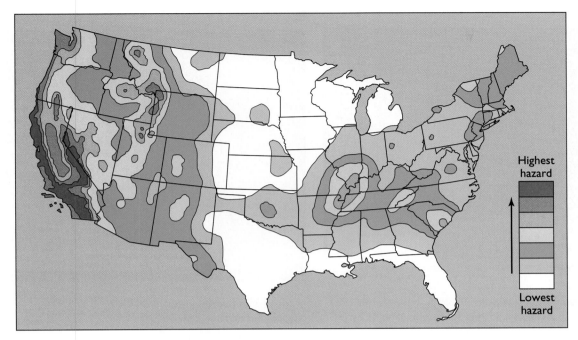

Figure 10.3 **Seismic hazard in the contiguous United States.** *(Courtesy of U.S. Geological Survey)*

Problem 10.6. Earthquake Magnitude and Level of Destruction. Imagine that Los Angeles experiences an earthquake of Richter magnitude 4 today, while Seattle experiences an earthquake of Richter magnitude 8. Referring to your text, answer these questions.

Q10.25. The most affected areas of Los Angeles probably will experience what level of damage?

Q10.26. The most affected areas of Seattle probably will experience what level of damage?

Q10.27. How many times larger in ground motion is the Seattle earthquake?

Problem 10.7. Tsunami: Warning Signs on Shore. Read your textbook's description of tsunami. Suppose you are at a harbor or along a coast and you notice that the sea level suddenly drops.

Q10.28. What would you expect to happen next?

Q10.29. If you want to save your life, what action should you take?

Problem 10.8. An Earthquake Leads to a Landslide. Study Figure 10.4 on the following page. The two photos show the same area, which is near Turnagain Heights in Anchorage, Alaska. Photo **A** was taken in 1961. Photo **B** was taken in 1964, a few months after a large quake struck Anchorage. The earthquake shook soil and sediment that overlies a wet layer of clay. Multiple landslides occurred as material slid toward the sea.

Q10.30. Measuring on the photos, what is the maximum north-south dimension of the slide area (in feet)? What is the maximum east-west dimension?

Q10.31. Using your answers to the preceding question, what is the maximum rough area for the slide (in square feet)?

Q10.32. Fortunately, most houses in Turnagain Heights were just beyond the slide area. However, not all structures escaped destruction. Count how many buildings appear to have been destroyed in the slide.

A

B

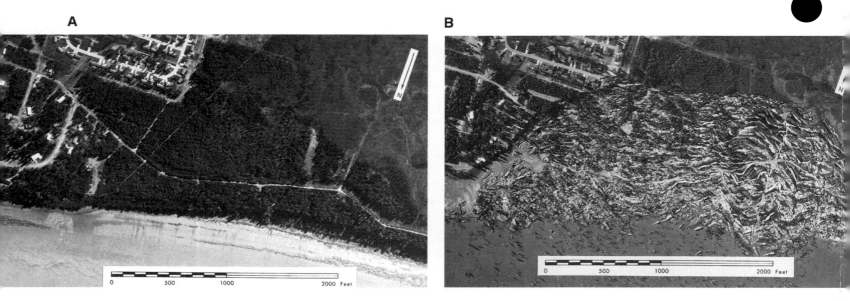

Figure 10.4 Turnagain Heights, Alaska. Photo **A** was taken in 1961, prior to the 1964 earthquake. Photo **B** was shot in 1964 following the quake. *(Photos by U.S. Geological Survey)*

Problem 10.9. Why Can You Still See the Lone Pine Fault? The fault shown in Figure 10.5 is quite old. It last generated earthquakes when rocks along the fault shifted in 1915. Often, a fault trace like this one is quickly buried under soil and vegetation. But the Lone Pine Fault is still very clear.

Q10.33. Why is the fault still visible?

Q10.34. If this fault trace were buried beneath thick soil, as it would be in most parts of the world, would it be less of a hazard to human beings? Explain.

Figure 10.5 Lone Pine Fault. *(Photo by U.S. Geological Survey)*

Figure 10.6   Ground following the Landers, California, earthquake of 1992. *(Photo by Jason Morgan)*

Problem 10.10. Seeing the Effects of a Fault. The photo in Figure 10.6 was taken after the Landers, California, earthquake of 1992, which was relatively minor.

Q10.35. From the evidence in the photo, was the movement along the fault vertical or horizontal?

Q10.36. Estimate in meters the degree of offset produced in this earthquake.

Q10.37. What kind of fault is shown?

Unit 10   Earthquakes

Last (Family) Name_____First Name_____

Instructor's Name_____Section_____Date_____

Q10.1, 10.2, 10.3, 10.4.

Table 10.1			
Wave velocities			
Spring	**Spring length when extended (units)**	**Round-trip compression wave travel time (seconds)**	**Round-trip shear wave travel time (seconds)**
Looser spring			
Tighter spring			

Q10.5. _____(show units)

Q10.6. _____(show units)

Q10.7. (circle one)   "P-wave"   "S-wave"

Q10.8. _____ seconds

Q10.9. _____ seconds

Q10.10. _____ seconds

Q10.11. (circle one)   yes   no

Explanation: _____

_____

Q10.12. _____

_____

Q10.13. _____

_____

Q10.14. _____

_____

Q10.15. _____

_____

Unit 10   Earthquakes

Last (Family) Name_____ First Name_____

Instructor's Name_____ Section_____Date_____

Q10.16.

Time gap between P-wave and S-wave	Distance from recording station to earthquake epicenter (kilometers)
Table 10.2	
Calculating the distance from seismometer to earthquake epicenters	
1 or 2 seconds	
1 minute	
2.5 minutes	
5 minutes	

Q10.17. _____(give unit)

Q10.18. _____(give unit)

Q10.19. (circle one)   yes   no

Explanation: _____

_____

_____

_____

Q10.20. _____

_____

_____

_____

Q10.21. (circle best answer for each city)

Seattle: 1   2   3   4   5   6   7

Los Angeles: 1   2   3   4   5   6   7

Chicago:	1	2	3	4	5	6	7
Miami:	1	2	3	4	5	6	7
New York:	1	2	3	4	5	6	7
Q10.22. (circle one)	1	2	3	4	5	6	7

Q10.23. Hometown: _____

Earthquake risk: _____

Q10.24. _____ times larger

Q10.25. (circle one)   great   serious   minor   none

Q10.26. (circle one)   great   serious   minor   none

Q10.27. _____ times larger

Q10.28. _____

Q10.29. _____

Q10.30. _____ feet north–south

_____ feet east–west

Q10.31. _____ square feet

Q10.32. about _____ buildings

Q10.33. _____

Q10.34. (check one)

_____ yes, less of a hazard

_____ no, at least as hazardous (if not more)

Explanation: _____

_____

_____

Q10.35. (circle one)   vertical   horizontal

Q10.36. _____ meters

Q10.37. (circle one)   normal   thrust or reverse   strike-slip or transform

Landslides, Slumps, and Creeps

Soil and rock sometimes move under the direct influence of gravity. Dramatic examples of such movement are **landslides, slumps,** and **rockfalls.** Similar processes—but significantly slower—include what geologists call **creep.**

Geologists refer to any movement of Earth materials due directly to the force of gravity as **mass wasting.** This unit introduces basic concepts important to your understanding of mass wasting. The ideas you encounter here will have direct application later in your life, especially if you purchase a house or piece of land.

Lab 11.1 Water and the Angle of Repose

In this experiment, you will measure the maximum slope at which sand grains are stable. This measurement is called the **angle of repose.** You will also see how different moisture conditions in the sand create dramatically different slope failures (movements).

Materials
- small aluminum pie pan
- well-rounded quartz sand, dry
- same kind of sand, quite damp
- standard protractor (or a large version drawn on cardboard)
- small scoop or spoon
- container of water

1. Clear all books and papers away from the pan (this experiment is a bit messy). Put the pie pan near the center of your table.

2. Slowly pour a stream of dry sand into the center of your pie pan. Continue slowly pouring as the sand fills the pan and builds above the rim. Stop pouring when the sand is overflowing the pan onto the table on all sides of the pan. At this point, you should have the maximum possible amount of sand heaped into the pie pan.

3. Avoid disturbing your sand pile. Carefully measure the maximum (*steepest*) slope of the dry sand. This slope is called the *angle of repose.* (**Hint:** Measure in degrees up from the horizontal.)

Q11.1. What is the angle of repose for the dry sand?

4. Add a tiny part of a spoonful of dry sand to the top of your pile and watch the sand grains move. Repeat, observing the movement again.

Q11.2. Describe the downward movement of the dry sand. Note especially whether the sand grains move individually or in large groups.

5. Put all your dry sand back into its container. Sweep your table clean. Put the pie pan back in the middle of the table.

6. Repeat the experiment using the damp sand. Place a pile of the damp sand in the pan. (**Hints:** You'll probably have to scoop or spoon—not pour—the damp sand into the pie pan. You can gently pack the pile together, but use only gentle pressure.)

Q11.3. What is the maximum slope angle you can make with the damp sand?

Q11.4. Describe the downward movement of the damp sand. Note whether the sand grains move individually or in larger groups.

7. Pour water onto your pile of damp sand, enough to saturate the sand.

Q11.5. Can the saturated sand maintain its steep slope angle?

Applying What You Just Experienced
The angle of natural slopes for different sediments varies according to several factors, including particle shape, degree of sorting, and water content.

Q11.6. Suppose a house is built on a slope made of sediment. The slope is slightly greater than the angle of repose. Is this house at risk from mass wasting?

Q11.7. Which sand condition—dry or damp—permits slopes with steeper angles?

Q11.8. Which sand condition—dry or damp—lends itself to the most dramatic, quickest, and therefore most dangerous style of slope failure?

Keep your results in mind when you consider buying a house. Examine the Earth materials in the neighborhood. If the materials become saturated with water from a heavy rainfall or flood, think what might happen to your house!

Lab 11.2 Particle Shape and the Angle of Repose

In this experiment, you will observe the effect of particle shape on the maximum stable slope of Earth materials.

Materials
- small, solid pie pan
- well-rounded gravel (pea gravel)
- angular gravel (construction gravel)
- standard protractor (or a large version drawn on cardboard)

Procedure
1. Clear all books and papers away from the pan. Put the empty pie pan near the center of your table.

2. Slowly pour rounded gravel into the center of your pan. Continue slowly pouring as the gravel fills the pan and builds above the rim. Stop pouring when it is overflowing the pan onto the table on all sides of the pan.

3. Avoid disturbing your gravel pile. Carefully measure the maximum (steepest) slope angle of the gravel. This is the gravel's angle of repose. (**Hint:** Measure in degrees up from the horizontal.)

Q11.9. What is the angle of repose for the rounded gravel?

4. Put all your rounded gravel back into its container. Sweep your table clean. Put the pie pan in the middle of the table.

5. Repeat the experiment using the angular gravel.

Q11.10. What is the angle of repose for the angular gravel?

See whether your labmates found a similar difference between angles of repose, based simply on particle shape.

Applying What You Have Just Experienced

Q11.11. Highway engineers and landscape architects often require that "fill" be added to fill a low or excavated area. Fill can be crushed rocks, river gravels, or soil. Considering the results of your experiment, why might engineers need to know the type of material that will be used as fill?

Problems

Problem 11.1. Basalt Face and Mass Wasting. Study Figure 11.1. Note the apron of loose rocks at the base of the cliff. This is what geologists call *talus*. This loose material has reached its present location due to the mass wasting process known as **rockfall.**

Q11.12. The slope of the loose rocks, as measured where they abut the cliff, is about how many degrees (up from the horizontal)?

Q11.13. If the vertical distance shown in the entire photograph is about 400 feet, then the rock pile (apron) is about how many feet high?

Problem 11.2. Deducing Past Mass Movement. Study Figure 11.2.

Q11.14. What two pieces of evidence indicate that this slope has moved in the past and is likely to be still slowly moving?

Figure 11.1   Rock face near Palouse Falls, Washington. *(Photo by E. K. Peters)*

Figure 11.2 **Wallowa Mountains, Oregon.** *(Photo by E. K. Peters)*

Problem 11.3. Stability of Italian Gravels. Study Figure 11.3.

Q11.15. What is the apparent angle of repose of the gravel pile?

Problem 11.4. Dramatic Mass Movement. Figure 11.4 on the following page shows Leaky Lake in Alaska.

Q11.16. What series of events or conditions might have led to the situation that formed this lake?

Q11.17. If you could take a field trip to Leaky Lake, what evidence would you seek to check the probability of the events you listed in the preceding question?

Figure 11.3 **Gravel pit in Italy.** *(Photo by Sheldon Judson)*

Figure 11.4 Area around Leaky Lake, Alaska. *(Photo by R. G. McGimsey, courtesy of U.S. Geological Survey)*

Problem 11.5. Dramatic Mass Movement. In 1925, near what is now Grand Teton National Park in Wyoming, a significant mass-wasting event occurred. It is shown in Figure 11.5.

Q11.18. What is this type of mass wasting called?

Q11.19. The photo was taken in 1997. How many years after the mass-wasting event was the photograph taken?

Q11.20. Do significant areas remain largely without vegetation years after the mass-wasting event?

Figure 11.6 is a geologic cross section of the Gros Ventre area.

Q11.21. From the cross section, what two special factors appear likely to have contributed to the mass-wasting event?

Figure 11.5 Gros Ventre near Teton National Park, Wyoming. *(Photo by L. Davis)*

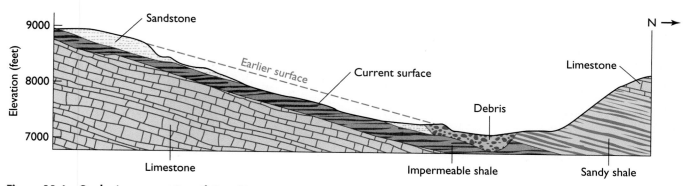

Figure 11.6   Geologic cross section of Gros Ventre, Wyoming. No vertical exaggeration. *(Drawn from data provided by the U.S. Geological Survey)*

Problem 11.6. Shoreline Mass Movement. Figure 11.7 shows the result of a mass-wasting event that occurred along the coast of California.

Q11.22. What type of mass wasting is evident?

Q11.23. What human-made structure has been disrupted?

Q11.24. Why are such structures often disrupted or destroyed by mass wasting?

Figure 11.7   Pacific Palisades, Los Angeles County, California. *(Photo by J. T. McGill, courtesy of U.S. Geological Survey)*

Figure 11.8 Agricultural fields in the kingdom of Bhutan. *(Photo by L. S. Hollister)*

Problem 11.7. Heroic Agricultural Efforts and Mass Movement. The kingdom of Bhutan is a small nation high in the Himalayan Mountains. Study Figure 11.8.

Q11.25. What structures have local farmers built to cope with the steep slopes of the kingdom?

Q11.26. Considering what you know about slopes and mass wasting, do you think these structures have to be repaired or rebuilt fairly often?

Problem 11.8. Sediment Produced by Mass Movement. Study the sedimentary material shown in Figure 11.9.

Q11.27. What evidence do you see in the sediment that mass wasting is likely to have been important in producing this material?

Q11.28. Imagine that you find a solid rock composed of particles like the large ones shown, cemented together with sand and calcite. In what environment do you think the rock formed?

Figure 11.9 Slope in Marble Canyon, Death Valley. *(Photo by L. Davis)*

Unit 11   Landslides, Slumps, and Creeps

Last (Family) Name_____ First Name_____

Instructor's Name_____ Section_____ Date_____

Q11.1. Dry sand angle of repose is _____ degrees.

Q11.2. Dry sand movement: _____

_____

_____

Q11.3. Damp sand slope angle is _____ degrees.

Q11.4. Damp sand movement: _____

_____

_____

Q11.5. (circle one)   yes   no, not fully

Q11.6. (circle one)   yes   no

Q11.7. (circle one)   dry   damp

Q11.8. (circle one)   dry   damp

Q11.9. Rounded gravel angle of repose is _____ degrees.

Q11.10. Angular gravel angle of repose is _____ degrees.

Q11.11. _____

_____

Unit 11 Landslides, Slumps, and Creeps

Last (Family) Name_____ First Name_____

Instructor's Name_____ Section_____ Date_____

Q11.12. _____ degrees

Q11.13. _____ feet

Q11.14.

1. _____

2. _____

Q11.15. _____ degrees

Q11.16.

1. _____

2. _____

3. _____

Q11.17. _____

_____

_____

Q11.18. _____

Q11.19. _____ years

Q11.20. (circle one)   yes   no

Q11.21.

1. _____

2. _____

Q11.22. _____

Q11.23. _____

Q11.24. _____

Q11.25. ───────────────────

Q11.26. (circle one)   yes   no

Q11.27. ──────────────────────

──────────────────────

Q11.28. ─────────────────

Mapping the Sun's Third Planet

Part V

This part of your lab manual presents two of a geologist's most valuable tools—topographic maps and geologic maps.

Topographic maps display Earth's surface features—valleys, mountains, canyons, hills, streams, and works of civilization. They depict Earth's three-dimensional surface on a two-dimensional sheet of paper. With a topographic map, you can accurately determine

- elevation of any point in the mapped area.
- steepness of a hiking trail, highway, or stream.
- land area drained by a stream or river network.
- distribution of water and vegetation in the mapped region.

Topographic maps are basic tools for geologists, civil engineers, land use planners, construction companies, military planners, archeologists, pilots, hikers, campers, hunters, homebuilders, and anyone who is interested in Earth's surface.

Geologic maps display the rock types exposed at Earth's surface, their structure, and their approximate ages. Geologic maps use different colors and patterns to represent each rock type. Engineers and geologists frequently use geologic maps. Some geologic maps are printed over topographic maps, providing information about both the landforms and the geology of the mapped region.

Topographic Maps: Revealing the Shape of the Land   Unit 12

The elevation and shape of Earth's surface result from a variety of geologic processes. Thus, every landform you see—hill, valley, plain, volcano, seashore—has a geologic history that determined its current shape. Topographic maps represent surface landforms with patterns of lines that precisely show elevations. Because of this, these maps help geologists to uncover the history of the landforms they study.

Lab 12.1   Getting Acquainted with Topographic Maps (Prelab)

This prelab walks you through two topographic maps to acquaint you with the basic information they contain.

Materials
- map in back pocket of this manual

Procedure

1. Remove the folded map from the pocket in the back of the manual. Unfold it carefully. The map is printed on both sides. You need the side that presents these two maps: Clark Fork, Idaho, and Palmyra, New York.

2. Let's roam the Clark Fork Quadrangle map. At first, this topographic map may look like a blur of lines. But once you study it and answer the following questions, it will become familiar—and quite interesting.

Q12.1. What is the name of the river (not its branches) that runs across the lower portion of the entire map?

Q12.2. What is the name of the large lake at the western edge of the map?

Q12.3. Where the river meets the lake, there are a number of large and small islands. What is the name of the only named island in the river?

Q12.4. The Clark Fork River branches into three forks where it meets the lake. Which fork is the narrowest overall?

Q12.5. What is the name of the national forest shown on this map?

Q12.6. Write the name of the mountain peak named for a feline carnivore that is found in the area.

Q12.7. What is the dashed line that goes to the top of Bee Top Mountain? (**Hint:** See the "Road Legend" on the map.)

Q12.8. Find a stream name on the map that indicates the presence of pesky insects in this area, at least during the summer months.

Q12.9. Several creeks drain into the Clark Fork River. In what general compass direction does Lightning Creek run?

Q12.10. The map is covered with a grid of fine red lines, forming squares about $2\frac{1}{2}$ inches on each side. Each square has a red number at its center. This grid is part of a system that can help you describe where a place is. The numbers repeat, so there are two number 1s, two number 2s, etc. What is the number of the red square that contains Howe Mountain (left central portion of map)?

Q12.11. What is the red number in the red square that contains the buildings of the Clark Fork Field Campus of the University of Idaho (southeastern corner of map)?

3. Now let's roam the Palmyra Quadrangle map. Look at the lower right margin of the map. Find the small "locator map" of the state of New York. The tiny black rectangle shows the area depicted by this topographic map. It takes hundreds of these little black rectangles to completely fill in the New York State outline. This means that hundreds of topographic maps like this one are required to depict the entire state of New York.

Q12.12. Where within New York State is the Palmyra Quadrangle?

Q12.13. What is the name of the canal that runs roughly east-west across the middle of the map?

Q12.14. What is the specific name of the stretch of canal that lies just northeast of the town of Port Gibson?

Q12.15. What school lies on Hyman Road in the northeast portion of the map?

Q12.16. What named hill lies just south of the town of Palmyra?

Q12.17. Portions of which two New York counties are shown on this map?

Q12.18. Which town is larger, Palmyra or Port Gibson?

Q12.19. Two railroads run across part of this map. What are their names? Draw the symbol used on the map to represent the railroads.

Q12.20. Find Huckleberry Swamp, northeast of the town of Palmyra. Draw the symbol used to show swampland.

4. Now look at the numerous curving brown lines on the map. These are **contour lines.** Each line represents a specific elevation above sea level. For example, in the lower left corner of the map, about 1 inch from the corner, is a brown contour line labeled "580." Every point on that line is 580 feet above sea level.

Q12.21. As you look around the map, are the brown contour lines roughly parallel to one another in any places?

Q12.22. Do the brown contour lines tend to cross over or intersect each other?

Q12.23. Do any of the brown contour lines run off the edge of the map?

Q12.24. Find Walton Hill, just north of Palmyra. Do the contour lines on the hill go around and form loops back on themselves?

Q12.25. Find Galloway Hill near the map's center. Do the contour lines on the hill go around and form loops back on themselves?

5. Now you have a basic acquaintance with topographic maps. Carefully fold your map and put it back in your lab manual. You will use this map repeatedly in lab.

Lab 12.2 Understanding Elevation and Topography by Looking at Shorelines

Topography means "to represent a location with lines," and that is exactly what a "topo" map does. It represents landforms through the use of contour lines. A good way to understand how this is done is to study shorelines.

A **shoreline** is the intersection where land and water meet. The elevation of the ocean's shoreline is called **sea level.** To help you picture these concepts, think of walking around the edge of an island, with your feet always right where the ground and the water meet. You would be walking along the shoreline at the same elevation all around the island, never rising nor falling. The elevation is the same everywhere because the force of gravity makes water level. This is why we usually call sea level "0 feet" and use sea level as a global reference point for elevations (example: "The top of Pike's Peak in Colorado is 14,110 feet above sea level").

As you will see in this experiment, understanding a shoreline is the key to understanding how the lines on topographic maps represent elevation.

Materials
- landforms made from modeling clay (inside water-proof containers)
- rigid plastic sheet (made of Plexiglas™ or equivalent)
- transparency ($8\frac{1}{2}''\times 11''$ sheet)
- marker or grease pencil for writing on the transparency
- water
- paper towels

Procedure

CAUTION ▶ Don't touch the modeling clay because you could deform it! Just study the model.

1. Note that one side of the container is marked N for north. That side should be away from you so that north for the container is "up" as you look down on the model. See Figure 12.1A.

2. At the southern end of your model (toward you) is the flat bottom of the container. This flat bottom represents the level surface of a lake. To the north of the lake are one or more landforms.

Q12.26. In Table 12.1 on the Lab Answer Sheet, describe the landforms you observe in your model. Use common words like "a lake and a long valley."

3. Let us say that the surface of the lake is at an elevation of 1000 feet above sea level. (We choose 1000 feet simply because it is a convenient number to start with.) We know that the lake's surface has the same elevation everywhere because water is naturally leveled by the pull of gravity. Thus, the top of the lake is a natural horizontal surface.

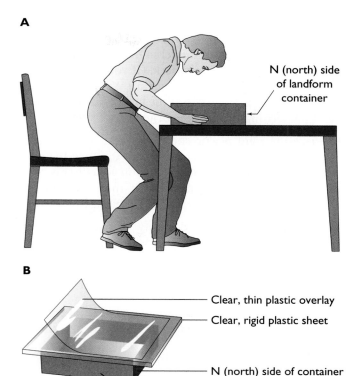

A

N (north) side of landform container

B

Clear, thin plastic overlay
Clear, rigid plastic sheet

N (north) side of container
Opaque plastic box

C

Tick marks

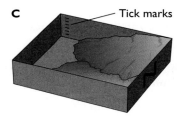

Figure 12.1   Clay landform model. A. Viewing position. **B.** Parts of the model box. **C.** General appearance of the clay model landform in box.

4. Place the rigid plastic sheet over your container.

5. Place the transparency on top of the rigid plastic sheet. (See Figure 12.1B.)

6. Looking *directly down* on the model, not from an angle, mark the corners of the model container on your transparency. Also draw a north arrow pointing toward the model's north end. (This way, if the transparency slides around, you can reorient it.)

7. Looking directly down on the model, trace the shoreline of the lake onto your transparency. This is a contour line, and it represents the intersection of the horizontal lake surface with the landforms in your model. Label the line on your transparency with the value we agreed to: "1000" (1000 feet elevation).

8. Look at the inside corners of your container and locate the evenly spaced tick marks. (See Figure 12.1C.) Slowly add water to your model so that the water surface just reaches the lowest tick mark. (If you add too much water, pour some off and start again.) We will call the elevation of the water surface at the lowest tick mark 1100 feet (100 feet higher than the original lake surface).

9. Place the rigid plastic sheet over your model again and align the transparency on top of it. Draw the new shoreline represented by the water you added. This contour line marks the intersection of the water surface with the landform in your model. Label the line "1100."

10. Remove the cover and slowly add water to your model so that the surface of the water just reaches the next tick mark. We will call the elevation of this water surface 1200 feet (200 feet higher than the original lake surface).

11. Place the rigid plastic over your model again and align the transparency on top of it. Draw the new shoreline represented by the water you've added. This contour line marks the intersection of the water surface with the landform in your model. Label the line "1200."

12. Continue this procedure, 100 feet at a time (one tick mark at a time), until you have filled your model to the top tick mark and have drawn all the shorelines on the transparency.

13. When done, write your name (and that of any lab partners) at the top of the transparency.

Q12.27. In Table 12.1, sketch the contour lines (from the transparency) that represent your model's landforms beside your description of each landform on the clay model.

14. Repeat the whole procedure with another landform model, noting the numbers of the models you use in Table 12.1.

Applying What You Just Experienced

Maps that show the shape of the land's surface must represent three-dimensional information on a two-dimensional piece of paper. Your transparency does exactly this by showing successive shorelines.

This is how topographic maps work. The "shorelines" are lines where horizontal planes of different elevations intersect a three-dimensional surface. The "shorelines" are called *contour lines*. Every point on a given contour line is at exactly the same elevation.

Q12.28. Contour lines that are close together indicate that the land has what kind of slope (steep or flat)?

Q12.29. Contour lines that are far apart indicate that the land has what kind of slope?

Q12.30. If you walk along any single contour line, will you be walking uphill, downhill, or on the level?

Q12.31. If you walk at right angles across many contour lines in a short amount of distance, will you be walking on a slope or on the level?

Q12.32. To get from one contour line to another next to it, what must you do?

Q12.33. Animals usually travel on the easiest route. They often form "cow paths" from repeatedly walking in the same place. On a hillside, do these cow paths usually run roughly parallel to the contour lines or across them at 90 degrees?

Lab 12.3 Reading Topographic Maps

This lab is a map adventure that helps you discover the wealth of information presented on topographic maps. The lab is in four parts:

> Part A   In the Margins of Topographic Maps
> Part B   How Contour Lines Depict Landforms
> Part C   Patterns in Contour Lines
> Part D   How Steep Is the Path Ahead?

Materials
- ruler marked in both inches and centimeters
- map in back pocket of manual
- key to map symbols (end of manual)
- optional: four-function calculator

Procedure
Remove the large map from the back pocket of your lab manual. Spread it out with the two topographic maps, Clark Fork and Palmyra, facing up.

Part A   In the Margins of Topographic Maps

The margins of topo maps contain lots of information. We'll begin at the upper left of each of your topo maps. This area shows the map's publisher, the U.S. Geological

Survey. (To see what topographic maps are available, visit *www.usgs.gov* and look under "products.")

The upper right corner shows the map quadrangle name, the state, and the map series.

- The map name comes from a prominent feature within the map's area—a community, stream, mountain, etc. (look for it on each map).

- "Series" indicates that this map is part of the series that covers a specific amount of area. For example, the Palmyra and Clark Fork Quadrangle maps are part of the 7.5-minute series, which means that each covers an area that measures 7.5 minutes of latitude by 7.5 minutes of longitude, or very roughly 70 square miles. (This is a very small area, allowing the maps to show great detail.)

The lower margin of a topo map gives technical information useful to geologists, engineers, and hikers. One of the most important pieces of information in the map's lower margin is its **scale.** Choosing the scale at which to draw a map is like choosing clothing that fits. Scale controls the size of the area a map can show, which means it also controls the amount of detail it can show. The easiest way to understand map scale may be to think of a scale-model car. Small scale-model cars sold in toy stores usually are ¹⁄₆₄ scale. This means that any part on the model is only ¹⁄₆₄ the size of the part on a real car. For example, a model of a 16-foot car is only ¹⁄₆₄ that long, or about 3 inches.

Map scales work exactly the same way, because a topo map is a scale model. In the case of the Palmyra and Clark Fork Quadrangles, the maps are 1/24,000 the size of the ground area they represent. This scale is shown in the lower margins of the maps as the ratio 1:24,000. This means that one unit of measure on the map = 24,000 of the same unit on the ground. For example:

1 inch on the map = 24,000 inches on the ground

1 foot on the map = 24,000 feet on the ground

1 dollar bill length on the map = 24,000 dollar bill lengths on the ground

Let's apply this information. With a ruler, measure the distance on the Palmyra Quadrangle from Huckleberry Swamp (northeast of the town of Palmyra) to School No. 1 on Schoolhouse Road. As you can see, it is almost exactly 1 foot. How far is this same distance in the real world?

You know that 1 foot on the map = 24,000 feet on the ground, so there is your answer: the distance is 24,000 feet. Let's convert this to miles: 24,000 feet on the

ground divided by 5,280 feet in a mile = about 4.5 miles from Huckleberry Swamp to Schoolhouse No. 1.

Q12.34. Using the same method, determine how far it is from the top of Galloway Hill (the hilltop is shown by an **X**) to the center of Galloway Cemetery, across the canal. Give your answer first in inches, then in feet.

Topo maps also include visual scales (near the bottom of the map) for determining horizontal distances. Note that the maps provide bar scales in miles, feet, and kilometers.

Be careful when you use the bar scales. Topo mapmakers often put the zero (0) point of the bar in the *middle* of the bar scale. Look at either map's mile bar scale. As you can see, from the 0 mark, the map shows you 1 mile to your left (divided into tenths) and 1 mile to your right.

Q12.35. Using the bar scale, determine how many inches on the map represent 7000 feet in the real world.

Q12.36. Using the bar scale, how many centimeters on the map represent 1 mile in the real world?

Q12.37. Using the bar scale, how many centimeters on the map represent 1 kilometer in the real world?

Part B   How Contour Lines Depict Landforms

Topo maps look like mazes of thin brown lines. These are the **contour lines** that represent elevation. This is the key distinguishing feature of a topo map.

Each contour line connects equal points of elevation, as you saw in Lab 12.1. The patterns of contour lines depict three-dimensional features like hills, mountains, and valleys on a two-dimensional map surface.

Let's go to an example on the Palmyra Quadrangle. Find Walton Hill, just north of the town of Palmyra. Confirm that the hill is shown by nearly 20 concentric rings or oval loops.

The **contour interval** is the vertical distance, or the difference in elevation, between two neighboring contour lines. As you can see on the map just below the bar scales, the contour interval for this map is 10 feet. This means that if you actually walked the ground represented on the map, walking from one contour line to the next contour line, you would either rise or drop 10 feet in elevation.

Q12.38. What is the contour interval of the Clark Fork map?

Q12.39. From this information alone, which of your two topo maps is likely to show greater relief (change in elevation)?

Note on both your maps that every fifth contour line is slightly darker and marked with an elevation value. These lines are called **index contours,** and they provide a reference for determining elevations.

Look again at Walton Hill on your Palmyra map. Looking at the contour lines that circle the hill, find the index contours marked 600 and 500. (The index contour between them is at 550 feet. It is not marked with a number because there is no room.) Note that the regular contour lines, each showing 10-foot intervals of elevation, roughly parallel these index contours.

Q12.40. What is the elevation of the highest contour line on Walton Hill?

Q12.41. Now look at the north slope and the south slope of Walton Hill. Based on what you know about contour lines, which slope is steeper overall? In other words, is it steeper to walk down from the top of Walton Hill toward the north or toward the south?

A few points on your Palmyra map have elevation marks of special significance. At these points, the U.S. Coast and Geodetic Survey or the U.S. Geological Survey has surveyed and placed permanent, physical elevation markers called *benchmarks* (BM). These markers are usually brass, iron, or concrete, and they are indicated on maps with an **X** or a small triangle Δ. Benchmarks are usually placed at prominent locations such as hilltops, road intersections, bridges, and so on.

Q12.42. What is the elevation of the benchmark at the top of Galloway Hill (center of map)?

Part C   Patterns in Contour Lines

Now let's consider patterns that contour lines make on maps and the six important rules that apply to contour lines.

Rule 1. All contour lines close to form a loop, no matter how big the loop may be. The loop may extend across an entire map or even several adjoining maps. This makes sense when you consider the example of an island. Its shoreline forms a single contour line of the same elevation that is a closed loop around the island, regardless of the island's size.

Look at the Clark Fork map. Note that the elevation of Lake Pend Oreille is given on the map as 2062 feet.

Q12.43. The shoreline around Derr Island lies at exactly what elevation?

Find the highest index contour line that runs around Antelope Mountain (southeastern corner of map). Note that it makes a roughly triangular loop that is entirely on this map.

Q12.44. Now find the 4000-foot contour that runs around Antelope Mountain. Does it make a closed loop on this map, or would you need the adjacent map to see the whole loop?

Rule 2. Contour lines never cross. (There is one rare exception: contour lines that represent an overhanging cliff must cross, so they are drawn as dotted lines.) Study the Clark Fork map and you will see that the contour lines may come very close to one another, but they never cross.

Q12.45. Are the contour lines near the top of Bee Top Mountain closer together than those near the top of Antelope Mountain?

Rule 3. Because contour lines connect points of equal elevation, they can never branch or divide. To confirm this, follow a few contour lines all around Howe Mountain (left center of map).
Rule 4. If a slope is gentle, contour lines are relatively far apart. If a slope is steep, contour lines are relatively close together. We discussed this in Lab 12.1 with reference to Walton Hill on the Palmyra map. Examine the spacing of contour lines on the Clark Fork map, and you'll see that the rule applies.

Q12.46. In the southeastern corner of the Clark Fork map, is the land steeper in the immediate area around the Lawrence Mine or the University of Idaho Clark Fork Field Campus?

Rule 5. In a valley, contour lines form a V that always points upstream, or up the valley. This is known as the "rule of Vs." You can see this very well at the stream in Webb Canyon on the eastern edge of the Clark Fork map. Notice where the stream occurs and how the contour lines make a V exactly where they intersect the stream.

Q12.47. What is the general compass direction for "uphill" in Webb Canyon?

Rule 6. Depressions without outlets have contour lines marked with hachures (ticks) pointing toward the center of the depression.

Q12.48. Locate Still Lake in the northeast corner of the map. How many contour lines around the lake should be marked with hachures? (The USGS omitted them.)

Use the rules to recognize landforms:

1. **Hills** and **mountains** are represented on a topo map by series of concentric circles or loops (Figure 12.2A).
2. **Depressions** are shown by concentric loops, at least one of which is marked by hachures (inward-pointing tick marks) (Figure 12.2A).
3. **Ridges** are shown by bends in contour lines that form elongate loops around hills or mountains (Figure 12.2B).
4. **Valleys** are represented by bends or Vs in contour lines (Figure 12.2B).
5. **Saddles** (also called "passes") between hilltops are represented as a gap between two different sets of concentric loops (Figure 12.2B).

Q12.49. Looking at the contour lines on the Clark Fork map, it's clear that Becker Draw (upper center of map) is an example of which of the above five landforms?

Q12.50. Look at the saddle between the two peaks of Howe Mountain (left center of map). What is the minimum elevation of the saddle?

Q12.51. What is the maximum elevation of the land in the saddle?

Q12.52. What landform rises to the northwest immediately across Lightning Creek from the town of Clark Fork?

Q12.53. Does Porcupine Lake (top center) lie in a closed depression?

Q12.54. Now look at the Clark Fork map as a whole. How would you describe its landforms overall?

Let us now return to the Palmyra map and consider its landforms.

Q12.55. There are numerous hills all across the map. What is their general shape?

Q12.56. When the Erie Canal was built, did the builders avoid the hills or cut through them?

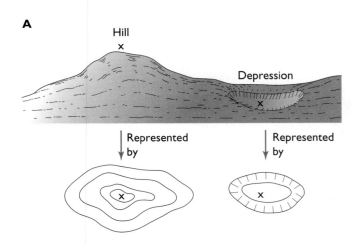

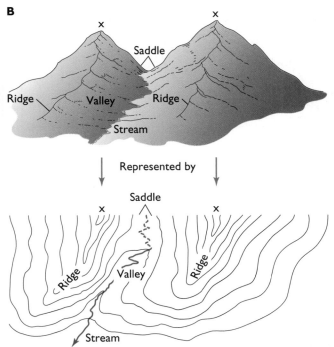

Figure 12.2 Topographic features.

Q12.57. Are there numerous or few closed depressions on the Palmyra map?

Now look at the Palmyra Quad as a whole.

Q12.58. How would you describe the map's landforms?

Part D   How Steep Is the Path Ahead?

Figure 12.3A shows a topographic map with a line drawn across it (**A–B**). This line shows the path we want to walk from point **A** to point **B**. How steep is this path? Do we travel up, down, or both? How far do we travel? You can easily answer such questions by drawing a **topographic profile.**

Procedure

1. Lay a strip of paper on the map so that the paper's edge follows line **A–B** (our walking path). See Figure 12.3B.

2. Mark a tick on the paper at every point where the edge of the paper crosses a contour line. Next to each tick mark, write the elevation of that contour line (Figure 12.3B).

3. Where you repeatedly cross the same contour line (such as the 200 line in Figure 12.3B), determine whether the land *between* those repeated intersections is higher or lower than the elevation of the line you cross.

4. Prepare a piece of graph paper as shown in Figure 12.3C. This will become your topographic profile. Label the ends of the profile A and B. Label the horizontal lines with regularly spaced elevation values.

5. Lay your paper with the tick marks and elevations on the graph paper as shown. At each tick mark, mark a dot at the same elevation on the graph paper.

6. Sketch the profile of the land from A to B by connecting the dots. Keep in mind where the profile goes higher or lower, and draw your lines accordingly. Your sketch is a picture of the "lay of the land" you would experience if you walked straight from A to B as shown on the map.

You now have constructed a topographic profile, which is a *vertical* "slice" of the land represented by the topo map. This allows you to view the land in profile, as if you cut away a piece of land and looked at it directly from the side, rather than from above.

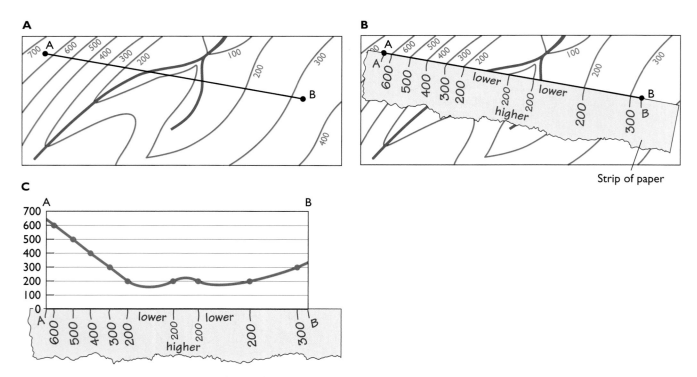

Figures 12.3 Constructing a topographic profile. A. Topographic map with profile line **A–B.**
B. Recording contours and elevations. **C.** Grid for topographic profile.

7. Now you will draw your own profile, following the procedure in steps 1–6. Find the point of your Clark Fork map where the creek in Becker Draw joins Spring Creek (above center of map). Mark that confluence with a dot. Imagine walking in a straight line from the top of Cougar Peak to the confluence you just marked.

Q12.59. To get a picture of what your walk would be like, draw the topographic profile, using the grid provided in Figure 12.4 on your Lab Answer Sheet. (**Note:** Use only the information conveyed by the index contour lines on the map.)

On your topographic profile, the vertical and horizontal scales are not the same. The horizontal scale is 1:24,000, just like the topographic map, but the vertical scale is 1:12,000. This means that the ups and downs you have drawn are exaggerated by a factor of two. The purpose is to make it easier to see the relief. (Topographic profiles can be drawn either without exaggeration to show the true slope of hills and valleys or with different vertical exaggerations to emphasize relief.)

Q12.60. If you converted your work to a profile without vertical exaggeration, what would its vertical scale become?

Unit 12   Topographic Maps

Last (Family) Name_____ First Name_____

Instructor's Name_____Section_____Date_____

Q12.1. _____ River

Q12.2. Lake _____

Q12.3. _____ Island

Q12.4. _____ Fork

Q12.5. _____ National Forest

Q12.6. _____

Q12.7. (check one)

_____ a stream

_____ an improved (paved) road

_____ a trail

Q12.8. _____ Creek

Q12.9. (check one)

_____ from the northeast toward the south

_____ from the northwest toward the east

_____ from the southeast toward the northwest

Q12.10. _____

Q12.11. _____

Q12.12. (check one):

_____ northeast corner

_____ southeast corner

_____ western part of the state

Q12.13. _____ Canal

Q12.14. _____

Q12.15. _____

Q12.16. _____ Hill

Q12.17. _____ and _____ Counties

Q12.18. (circle one)   Palmyra   Port Gibson

Q12.19. _____ Railroad and _____ Railroad

Draw railroad symbol:

Q12.20. Draw swampland symbol:

Q12.21. (check one)

_____ yes, roughly parallel

_____ no, not at all

Q12.22. (circle one)   yes   no

Q12.23. (circle one)   yes   no

Q12.24. (circle one)   yes   no

Q12.25. (circle one)   yes   no

Unit 12   Topographic Maps

Last (Family) Name_____ First Name_____

Instructor's Name_____ Section_____ Date_____

Q12.26, 12.27.

Table 12.1		
Model landforms		
Model Number	**Describe landforms in your model (valley, plain, canyon, hill, ridge, cliff, etc.).**	**Sketch the contour lines from your transparency that correspond to each feature.**

Q12.28. _____

Q12.29. _____

Q12.30. (circle one)   uphill   downhill   on the level

Q12.31. (check one)

_____ on a slope

_____ on the level

Q12.32. (check one)

_____ change elevation

_____ use your compass

Q12.33. (check one)

_____ roughly parallel to the contour lines

_____ across the contour lines at 90 degrees

Q12.34. _____ inches _____ feet

Q12.35. _____ inches on the map

Q12.36. _____ cm on the map

Q12.37. _____ cm on the map

Q12.38. _____ feet

Q12.39. _____ map

Q12.40. _____ feet

Q12.41. (check one)

_____ steeper toward the north

_____ steeper toward the south

Q12.42. _____ feet

Q12.43. _____ feet

Q12.44. (check one)

_____ It makes a closed loop on the Clark Fork map.

_____ It runs off the edge of this map (and probably loops back on the adjacent map).

Q12.45. (circle one)   yes   no

Q12.46. (check one)

_____ Lawrence Mine

_____ Clark Fork Field Campus

Q12.47. (circle one)   NW   NE   SW   SE

Q12.48. (circle one)   0   1   2   3

Q12.49. _____

Q12.50. just greater than_____ feet

Q12.51. just below_____ feet

Q12.52. _____

Q12.53. (circle one)   yes   no

Q12.54. (check one)

_____ broad plain with numerous but small elongate hills

_____ sharply mountainous region with several major, nearly flat-bottomed valleys

_____ deep and long canyon bottom

Q12.55. (check one)

_____ well-rounded, equal lengths

_____ elongated in a general east-west direction

_____ elongated in a generally north-south direction

Q12.56. (check one)

_____ avoided the hills

_____ cut through them

Q12.57. (circle one)   numerous   few

Q12.58. (check one)

_____ broad plain with numerous but small, elongate hills

_____ sharply mountainous region with several major, nearly flat-bottomed valleys

_____ deep and long canyon bottom

Q12.59.

Figures 12.4 Grid for constructing a topographic profile.

Scale 1:12,000

Confluence
of streams

Scale 1:24,000

Cougar
Peak

6400 ft

5400 ft

4400 ft

3400 ft

2400 ft

Elevation (feet)

Q12.60. 1: _____

Rock Deformation and Geologic Maps — Unit 13

Below the surface of our planet, rocks experience substantial forces. These forces include **compression** ($\rightarrow\leftarrow$), **extension** ($\leftarrow\rightarrow$), and **shearing** ($\leftrightarrows$). All of these forces deform rock. Not only can the forces bend rocks (creating a **fold**), they can even break them (called a **fault**). These forces help create the structure we observe in rocks.

Geologic maps show rocks that are exposed at Earth's surface or just below the surface. Geologic maps usually ignore the surface soil and show only the rocks that lie beneath loose surface debris. Geologic maps show each type of rock and, where useful, a rock layer's orientation. The maps also show the rock structures that result from deformation, such as folds and faults.

Lab 13.1 Getting Acquainted with the Pine Grove, Pennsylvania, Geologic Map (Prelab)

Your first encounter with a geologic map is like a visit to a strange neighborhood. The best way to get acquainted is to roam the neighborhood for awhile. That is exactly what you do in this lab.

Materials
- Pine Grove Geologic Map (in map pocket at back of book)

Procedure
1. Remove the map from the pocket in the back of this manual. Carefully unfold it. You need the side with the Pine Grove Geologic Map. Study the map. Don't panic! It looks much more complex than it is. The geologic information on the map is superimposed on a topographic map of the area. Everything you learned about reading topographic maps in the preceding unit will help you to interpret the Pine Grove Geologic Map. Here are some questions to get you acquainted with this map.

Q13.1. What is the scale of the map, given as a ratio?

Q13.2. This means that 1 inch on the map represents how many inches on the ground?

Q13.3. What is the contour interval of the map?

Q13.4. A "run" is a small stream. Find Bear Hole Run in the southwestern corner of the map. Note the many contour lines that cross the stream. Where they cross, the contour lines form Vs. Which way do the Vs of the contour lines point?

Q13.5. What is the general compass direction in which Fishing Creek flows?

Q13.6. If you drove from the town of Suedberg to the outskirts of Pine Grove, you would travel in which direction?

Q13.7. What is the approximate elevation of Pine Grove at the point where highway 443 intersects with highway 125?

Q13.8. The town of Pine Grove has set up a TV cable receiving station (extreme southeastern corner of the map). Such stations must be at high points to receive strong signals. What is the elevation of the Pine Grove receiving station? (**Hint:** The station is shown as a light gray box. The dark black dots and lines in the area are unrelated.)

Q13.9. How much higher is the receiving station than downtown Pine Grove?

2. Now that you are more at home with the map, we will look at the geology it portrays. Examine the colors on the map (which show the rock units) and compare them with the color key (to right of map).

Q13.10. What is the name of the rock formation beneath the Pine Grove TV Receiving Station?

Q13.11. During what geologic period did this rock unit form?

Lab 13.2 Deformation of Sedimentary Rock Beds

When sediment settles out of water, the result usually is flat, horizontal layers. This commonly happens in lakes and oceans. If these sediment layers become lithified into rock, the rock layers also lie flat and horizontal, at least initially. Thus, sedimentary rocks begin their existence as flat-lying structures. Later, Earth's forces can deform (bend, fold, break) the rock layers into a variety of structures. In this exercise, we will use sheets of paper to represent four thin sedimentary beds.

Materials
- 4 sheets of heavy paper (such as cover stock), each a different color

Procedure

1. Imagine that each of the paper colors represents a different type of sedimentary rock. (For example, a yellow sheet could represent a bed of limestone, blue could represent sandstone, and so on.)

2. Stack the sheets in any order. Position the stack so that the longer side (11-inch) is toward you. In this simple model, the oldest rock bed is at the bottom (it was deposited first) and the youngest bed is on top.

3. Place your fingers against the left and right edges of the stack. Push downward to keep these edges on the table and push your hands toward each other so that the papers bow upward in the middle.

Q13.12. What type of force are you applying to the papers?

Q13.13. In this simple model, the stress you have applied has produced what common geologic structure?

Q13.14. What is the orientation of the axis of the structure you have made?

Q13.15. Does this structure show what geologists call plunge?

Q13.16. In the structure you have made, is the oldest rock bed still beneath the younger beds of rock?

4. Repeat step 3. In addition, tilt upward the edge of the paper that is nearest your face, keeping everything else the same. Look straight down the axis of the structure you have made.

Q13.17. In what direction does the structure now plunge?

Q13.18. Is the oldest bed of rock still beneath the younger beds?

5. Straighten the sheets so that their edges align. Place them flat on the table with the longer side (11-inch) toward you. Anchor one edge of the stack by holding it down firmly with your hand.

6. Using your other hand, pick up the other edge of the stack and push the stack so it folds over itself to form an S shape (viewed end-on).

7. On the topmost limb of your S, note that the oldest bed is still beneath the younger beds.

Q13.19. *In the middle portion* of the S, is this still true?

Q13.20. What is the name of this type of fold?

8. Position the stack flat on the table so that the longer side (11-inch) is toward you. Tap the right-hand edge of the stack until the four sheets form little stair steps. (The topmost paper will be a little more to the left than the sheet below it, and that sheet a little more to the left than the one below it, and so on down the stack.)

Q13.21. What type of deformation have you produced?

Applying What You Just Experienced
Sedimentary beds, because they are in layers, clearly show geologic structures in ways that, for example, a large block of granite cannot.

Q13.22. What is the name for the upward-bent, A-shaped portion of a fold?

Q13.23. Why can we say that overturned folds literally "turn the world upside down"?

Lab 13.3   Types of Deformation

When Earth materials undergo **stress,** the result is **strain** or **deformation.** In this experiment, you will discover the different conditions necessary to produce three different types of strain.

Materials
- Silly Putty™ or equivalent
- hammer
- thin block of wood for absorbing hammer blows
- plastic bags: sandwich size, zip-lock type

Procedure

1. The three major types of deformation that occur in Earth materials are **elastic, ductile,** and **brittle.** (Please review these deformation styles in your textbook to anchor the concepts clearly in your mind.)

2. When you have a clear concept of elastic deformation, experiment with your Silly Putty until you find a way to show elastic behavior.

Q13.24. In Table 13.1 on your Lab Answer Sheet, describe the procedure you used to show elastic deformation.

3. When you have a clear concept of ductile deformation, find a way to show this type of strain in the Silly Putty.

Q13.25. In Table 13.1, describe the procedure you used to show ductile deformation.

4. Put your Silly Putty into a plastic bag. Using the hammer and block of wood, produce brittle deformation in the Silly Putty.

Q13.26. In Table 13.1, describe the procedure you used to cause brittle deformation.

5. Re-form the Silly Putty into a single ball and return it to its container.

6. Two major variables regarding stress are (1) how *rapidly* the stress is applied and (2) how *great* the stress is.

Q13.27. In Table 13.1, show the rate at which you applied stress in each experiment: 1 = fastest, 2 = intermediate, and 3 = slowest.

Q13.28. In the last column of Table 13.1, show the magnitude of the stress you applied in each experiment: 1 = greatest, 2 = intermediate, and 3 = least.

Applying What You Just Experienced

Stress produces **strain** in rocks. **Strain** is the actual result of stress. Strain may vary from a slight warping of a sedimentary bed and complex flow textures in a metamorphic rock to the rupture of solid rocks along a fault.

Q13.29. Strain varies according to what two major variables related to stress?

Q13.30. If you and your lab partners used geologic terminology to describe your reaction to an upcoming exam, which would you say?

Q13.31. Name a geologic environment or process that is known for producing ductile deformation. (***Note:*** Several answers are possible.)

Lab 13.4 Common Geologic Structures in Three Dimensions

Geologists often have to visualize geologic structures in three dimensions (3-D) to understand how rocks got into their present position. They usually have to work from the limited clues they see on Earth's surface (or on geologic maps that depict the limited clues).

This exercise helps you visualize geologic structures in three dimensions. You will do so like a real geologist does, working from two-dimensional (flat) information that is drawn on the top of a box. The box is a model of a portion of Earth's rocks.

Materials
- several forms titled "What's below the Surface?"
- numbered boxes with geologic map information on top and sides marked "Face **A**" and "Face **B**"
- colored pencils
- protractor
- ruler

Procedure

1. Your instructor will give you several forms titled "What's below the Surface?" and a box representing rocks at and just below Earth's surface.

2. Complete the information at the top of a "What's below the Surface?" form.

3. Study the box. The top represents a small area of Earth's surface (perhaps a few square miles). The land it represents is a flat plain. On the top you will see rock symbols and names with strike-and-dip information (explained in Lab 13.5). These form a geologic "map" of the surface at this location.

4. Using what you know about rock formation and the geometry of different rock types, your task is to infer the most probable arrangement of rocks beneath the surface.

5. On the "What's below the Surface?" form, at "Face **A**," sketch in the rock names and symbols of face **A** on your box.

6. At "Face **B**" on the sheet, do the same for face **B** on your box.

7. At "Name of structure" on the sheet, write the complete name or description of the geologic structure you have drawn.

8. When you are finished with the first box, pass it on to your labmates and repeat the steps with another box.

9. When you have finished all the boxes, put your sheets in numerical order and staple them together.

Applying What You Just Experienced

This exercise illustrates the conservative way geologists reason. Like all scientists, geologists deal with observable, measurable reality; they avoid speculation. Scientists call this the principle of simplicity or economy. It's also known as "Ockham's razor," for William of Ockham, a medieval philosopher who believed that the simplest explanations are best. A modern version of this concept is the KISS principle (Keep It Simple, Stupid).

Q13.32. Imagine that you are surrounded by flat-lying limestone. How likely is it that beneath the limestone there is a gold-rich quartz vein? Using Ockham's razor, explain.

Lab 13.5 Introduction to Geologic Maps: Pine Grove, Pennsylvania, Geologic Map

Geologic maps show the rocks at Earth's surface and near-surface, not deep underground. They depict the area where each rock unit is exposed, how each rock unit is oriented, and where faulting occurs.

Notice that geologic maps are highly colorful. Dozens of different rock formations may appear at the surface within the mapped area, so numerous colors and patterns are needed to clearly differentiate each rock type. Note that the colored areas are separated by thin lines. These are called **contact lines** because they represent the point where the two rock types make contact.

Rock orientation is shown by small **strike-and-dip** symbols (⊤) within the colored area for a given rock. These symbols indicate whether a bed is horizontal or tilted. **Faults** and the **axes of folds** are shown as dark lines.

Many geologic maps also show topography because it is very helpful to relate the rock units to the shape of the land surface. Maps like the Pine Grove map, which combines geologic observations and topographic information, can look quite complex. But, with a little time and patience, you can read them easily enough.

A Quick Guide to Reading Geologic Maps

Strike and Dip. The orientation of sedimentary rock beds is shown on geologic maps by small strike-and-dip symbols. There are three types of symbols:

⊕ = perfectly horizontal bed

+ = perfectly vertical sedimentary bed (very rare)

⊤ = most sedimentary beds, oriented somewhere between exactly horizontal and exactly vertical. The *strike* is shown by the long line, the *dip* by the short line. Figure 13.1 shows how the system works.

1. Consider the sandstone bed ①. This sandstone was deposited horizontally, of course, but later was tilted to the angle you see in the figure. The original surfaces of the sandstone layer form its **bedding planes.**

2. We need a horizontal plane for reference, so we add one, which we can think of as the surface of a lake ②.

3. Notice the line where the horizontal plane intersects the rock's bedding plane (the red arrow). This line is

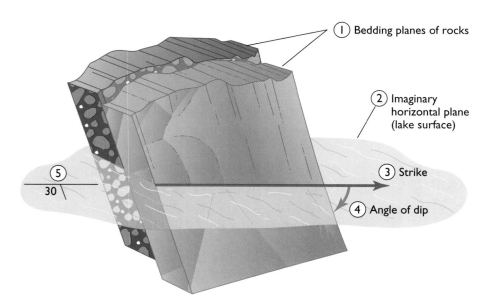

Figure 13.1 **How strike and dip symbols show rock orientation on geologic maps.**

called the **strike** ③. The strike line has a compass direction on Earth's surface, so we can describe the strike of the sandstone bed as a direction (like northeast).

4. Next we indicate how steeply the sedimentary bed dips relative to the horizontal plane. The angle of **dip** is measured *downward* from the horizontal plane (lake surface), as shown by the green arrow ④. This angle is expressed in degrees (for example, 30°).

5. Knowing the strike and the dip of the sedimentary bed, we can describe its three-dimensional orientation. Geologists record this strike-and-dip information directly on a geologic map using the strike-and-dip symbol ⑤. The longer line is the strike direction, the shorter line is the dip direction, and the number is the dip angle in degrees (down from the horizontal).

Faults. Faults—breaks in rock layers—are shown on a geologic map by black lines (Figure 13.2). The lines are solid where the fault is clearly evident. The lines are dashed where the fault is not clearly visible at the surface and must be inferred by a geologist.

Most faults are roughly planes. Thus, if the orientation of a fault plane can be measured, we can show its strike (direction) and dip (angle down from horizontal). Alternatively, we may only be able to show which side of the fault moved upward or downward relative the other side.

Thrust faults are shown as lines with triangular teeth. The teeth point toward the upper fault block (often called the "hanging wall").

Folds. In sedimentary rocks, the axes of folds are also shown on geologic maps as lines (Figure 13.3 on the next page). An *anticline* is drawn as a long line straddled by a two-headed arrow in which the arrowheads point away from one another. A *syncline* is drawn the same way with the arrowheads pointing toward one another.

If an anticline or syncline is not horizontal—in other words, if it *plunges*—the direction of plunge is indicated by an arrowhead on the end of the line.

In rare cases, where sedimentary beds have been fully overturned by folding, a U-shaped symbol is used to give the strike and dip of the bed.

Pine Grove Map

Answer the following questions about your Pine Grove map.

Q13.33. Does the Martinsburg Shale appear *at the surface* in this quadrangle?

Q13.34. Looking at the two geologic cross sections toward the left on the map, does the Martinsburg Shale appear *below the surface?*

Q13.35. Why did geologists sketch the Martinsburg Shale as they did?

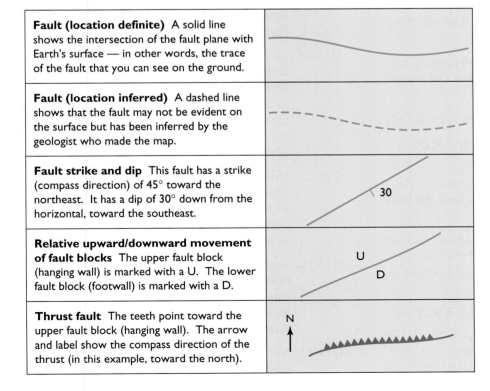

Fault (location definite) A solid line shows the intersection of the fault plane with Earth's surface — in other words, the trace of the fault that you can see on the ground.	
Fault (location inferred) A dashed line shows that the fault may not be evident on the surface but has been inferred by the geologist who made the map.	
Fault strike and dip This fault has a strike (compass direction) of 45° toward the northeast. It has a dip of 30° down from the horizontal, toward the southeast.	30
Relative upward/downward movement of fault blocks The upper fault block (hanging wall) is marked with a U. The lower fault block (footwall) is marked with a D.	U D
Thrust fault The teeth point toward the upper fault block (hanging wall). The arrow and label show the compass direction of the thrust (in this example, toward the north).	N

Figure 13.2 How faults are shown on geologic maps.

Anticlinal crests The long line shows the trace of the anticline. It is dashed where inferred. The two-headed arrow with the arrowheads pointing away from one another indicates that the structure is an anticline.	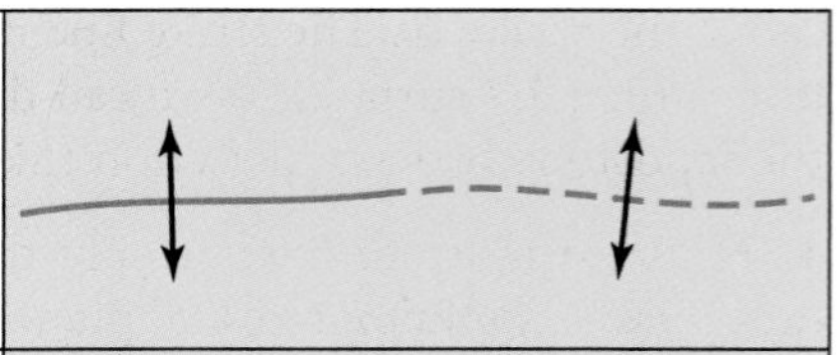
Synclinal trough The long line shows the trace of the syncline. It is dashed where inferred. The two-headed arrow with the arrowheads pointing toward one another indicates that the structure is a syncline.	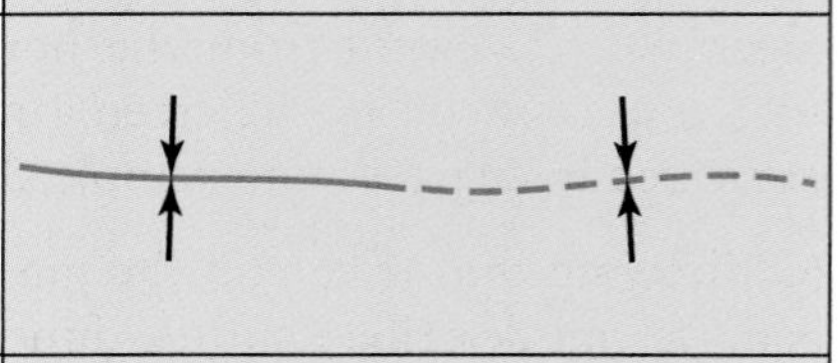
Plunging anticline Where the anticline is not horizontal but plunges, the direction of plunge is shown by the arrowhead — in this case, toward the west.	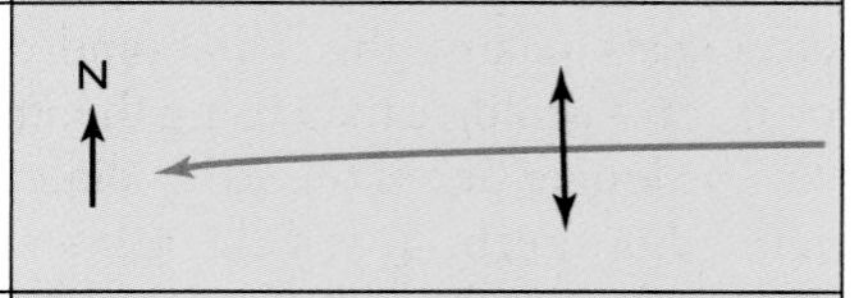
Plunging syncline Where the syncline is not horizontal but plunges, the direction of plunge is shown by the arrowhead — in this case, toward the east.	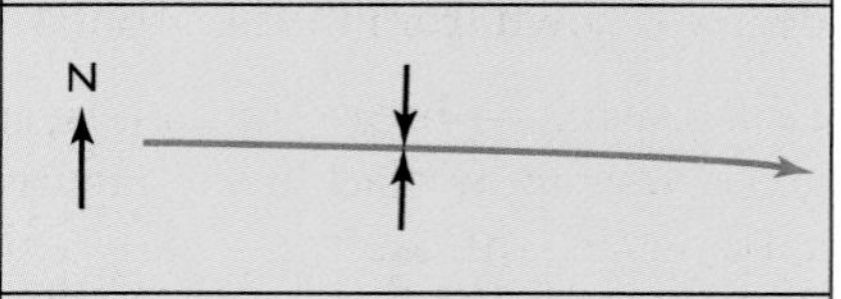
Overturned fold The U-shaped symbol drawn across the line of strike indicates that the fold has overturned the beds. The number shows that they are dipping 70° to the west.	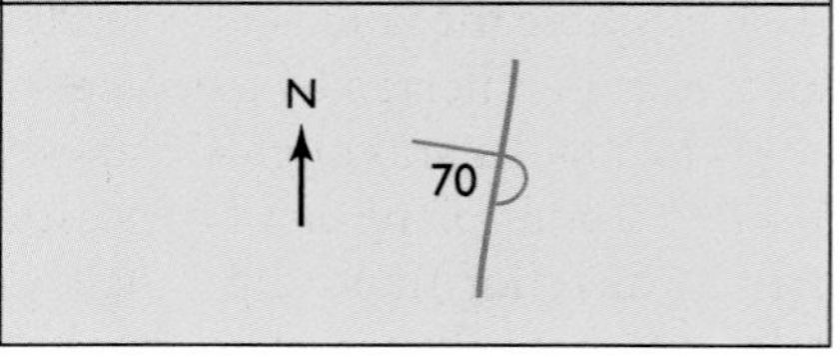

Figure 13.3 How folds are shown on geologic maps.

Q13.36. Looking back at your map, along what features are the **Qal** deposits located?

Q13.37. During what unit of geologic time was the **Qal** material deposited?

Q13.38. Considering your answers to the preceding two questions, is it probable that **Qal** material is still being deposited?

Q13.39. Along what features are **Qt** deposits located?

Q13.40. What kind of material makes up the **Qt** deposits?

Q13.41. Looking at the cross sections drawn on the left side of the map, how thick have the beds of **Qal** and **Qt** been drawn?

Q13.42. Now consider the faults of this area as shown on both the cross sections and the map. Which type of fault is more common in this quadrangle?

Q13.43. What kind of forces are associated with such faults?

Q13.44. What kind of forces are associated with the folding of rocks into anticlines and synclines?

Q13.45. From these observations, what would you say has been the general tectonic setting of this area at some point since the Pennsylvanian Period?

Q13.46. Look at the belt of Bloomsburg Red Beds that is shown on the lower portion of the map. Circle all eight strike-and-dip symbols shown in these rocks. What is the name of the structural feature on the map that accounts for the sharp difference in orientation shown by two of the strike-and-dip symbols?

Q13.47. Find the location of the structure in your answer to the preceding question on the cross section drawn from **A** to **A′**. Name the four beds on the cross section that reflect this structure.

Q13.48. Study the "Explanation" on the map to see how major coal beds in this area are indicated. Next, study the map. During what geologic period was virtually all of the coal of this area deposited?

Q13.49. Now study the coal beds as shown in the two geologic cross sections drawn toward the left on the map. What general type of mining do the coal beds in this region lend themselves to?

Q13.50. What structural features account for your answer to the preceding question?

Problems

Problem 13.1. Rock Contacts and Topography. Figure 13.4 shows four sedimentary beds and how they crop out at the surface.

Q13.51. What is the general relationship between the contour lines and the rock contacts?

Q13.52. Although there are no strike-and-dip symbols on this map, what would you say is the orientation of the sedimentary beds?

Q13.53. What is the elevation of the bottom plane of the fine sandstone bed?

Q13.54. What is the approximate elevation of the top plane of the fine sandstone bed?

Problem 13.2. Geology of Tennessee's Swan Island Quadrangle. Study the geologic map of part of Tennessee shown in Figure 13.5 on the following page. The rocks in this area are sedimentary beds, with the older rocks exposed toward the northwest and the younger rocks appearing on the map toward the southeast.

Q13.55. Do the contour lines and the rock contact lines always parallel each other on this map?

Q13.56. From that observation alone, is it likely that the sedimentary beds in this area are flat-lying?

Q13.57. Refer to the strike-and-dip symbols shown on the map. What is a generalized direction for the strike of all of these beds?

Q13.58. What is the range of values for the dip of the beds?

Q13.59. What symbol on the map represents loose sediment deposited very recently, geologically speaking? What does the symbol stand for?

Q13.60. Along what type of water feature do these loose sediments occur?

Problem 13.3. Completing Geologic Block Diagrams. Figure 13.11 on your Problem Answer Sheet shows four incomplete block diagrams.

Q13.61. In Figure 13.11A and B, sketch in the rocks on the surface and color them appropriately. Then label the geologic structure shown in each drawing.

Q13.62. In Figure 13.11C, sketch in the rocks on both blank sides and color them appropriately.

Q13.63. In Figure 13.11D, sketch in the rocks on the surface and color them appropriately.

Problem 13.4. Shear Deformities in Trilobites. Trilobites were sea creatures of the Paleozoic era. When rocks containing trilobite fossils are deformed by shear stresses, the trilobite fossils deform, too.

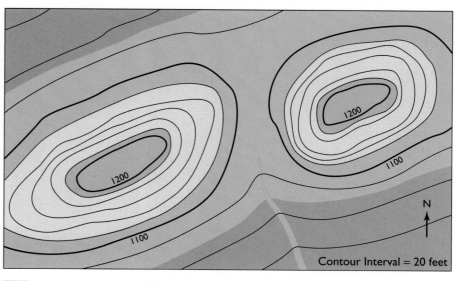

Figure 13.4 Four sedimentary beds and how they crop out at the surface.

Coarse sandstone   Fine sandstone   Shale   Limestone

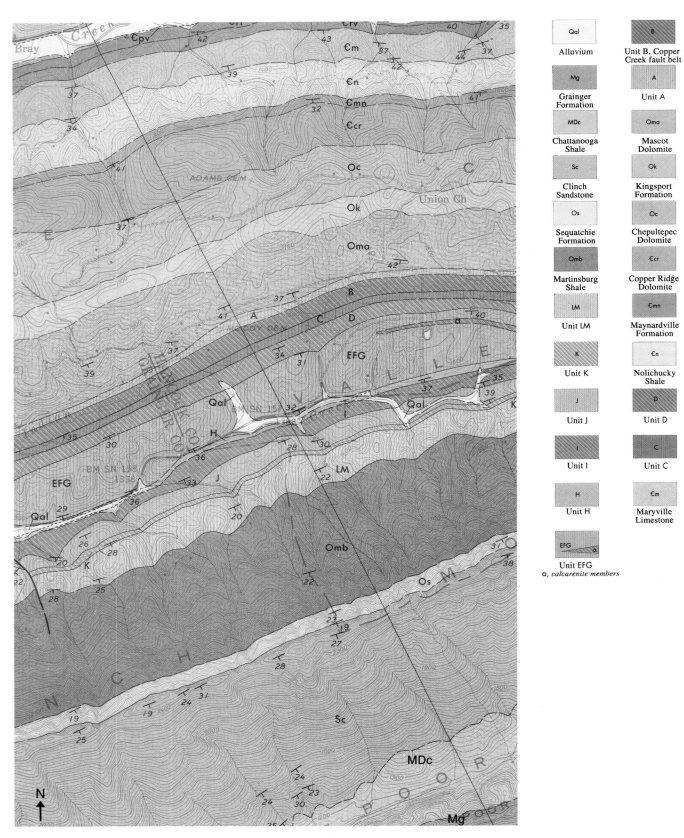

Figure 13.5 **Geologic map of part of the Swan Island Quadrangle, Tennessee.** 1:24,000 scale, contour interval = 20 feet. *(U.S. Geological Survey)*

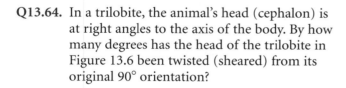

Figure 13.6 Trilobite showing the strain. *(Photo by Kathy Baldwin)*

Figure 13.7 Trilobites showing the strain. *(Photo by Kathy Baldwin)*

Q13.64. In a trilobite, the animal's head (cephalon) is at right angles to the axis of the body. By how many degrees has the head of the trilobite in Figure 13.6 been twisted (sheared) from its original 90° orientation?

Q13.65. Study the two trilobites in Figure 13.7. Geologists have good reason to think both animals were members of the same species and thus had the same body shape. What is the width-to-length ratio of the trilobite to the right?

Q13.66. What is the width-to-length ratio of the trilobite to the left?

Q13.67. Why do these two ratio values differ greatly?

Problem 13.5. Identifying Structures in Wyoming. Study the aerial photograph (Figure 13.8).

Q13.68. If you had no information about which way the beds are dipping, what two structures could the photo represent?

Q13.69. Notice that a single strike-and-dip symbol has been added to the photo (upper left area). Knowing this single dip direction, you can identify this structure. What is it?

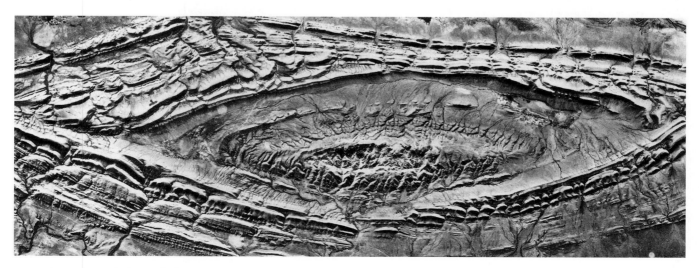

Figure 13.8 Aerial photo of Sheep Mountain, Wyoming. *(Photo courtesy of U.S. Geological Survey)*

Figure 13.9   Structure seen from Space Shuttle.
(Photo courtesy of NASA)

Problem 13.6. Structure Seen from Space. The Space Shuttle photo in Figure 13.9 shows the Sahara Desert in Mauritania, Africa. Geologists on the ground have determined that the rocks at the center of the structure are older than the rocks around its edges.

Q13.70. Knowing this, what is the geologic name for the structure?

Problem 13.7. Structure and Topography. Topographic high and low points are not generally determined by the structures within rocks. Figure 13.10 shows the top of a mountain through which a highway has been constructed in western Maryland.

Q13.71. What is the name of the geologic structure you see in the roadcut?

Q13.72. If the road runs roughly east-west, what is the rough compass direction of the axis of the structure in the rocks?

Figure 13.10   Sideling Hill, Maryland. *(Photo courtesy of Maryland Geological Survey)*

Unit 13   Rock Deformation and Geologic Maps

Last (Family) Name_____ First Name_____

Instructor's Name_____ Section_____ Date_____

Q13.1. 1: _____

Q13.2. _____ inches

Q13.3. _____ feet

Q13.4. (circle one)   uphill   downhill   sideways

Q13.5. (circle one)   NW   SW   due east   due west

Q13.6. (check one)

_____ due north

_____ a little north of east

_____ due south

_____ westerly

Q13.7. _____ feet

Q13.8. _____ feet

Q13.9. _____ feet

Q13.10. _____

Q13.11. (circle one)   Cambrian   Ordovician   Silurian   Devonian

Unit 13   Rock Deformation and Geologic Maps

Last (Family) Name_____First Name_____

Instructor's Name_____Section_____Date_____

Q13.12. (circle one)   compressional   extensional   shear

Q13.13. (circle one)   syncline   alkaline   anticline

Q13.14. (check one)

_____ running to my right and left

_____ running toward me/away from me

Q13.15.   (circle one)   yes   no

Q13.16.   (circle one)   yes   no

Q13.17. (check one)

_____ to my left

_____ to my right

_____ into the tabletop

Q13.18. (circle one)   yes   no

Q13.19. (check one)

_____ Yes, the oldest bed is still on the bottom

_____ No, the oldest bed is now on top

_____ Not enough information to tell

Q13.20. _____ fold

Q13.21. (circle one)   compressional   extensional   shear

Q13.22. (circle one)   anticline   syncline   monocline

Q13.23. _____

Q13.24, 13.25, 13.26, 13.27, 13.28.

Table 13.1

Three types of deformation

Deformation type	How did you produce this deformation with the Silly Putty?	Rate at which you applied stress (1 = fastest, 3 = slowest)	Magnitude of stress (1 = greatest, 3 = least)
Elastic			
Ductile			
Brittle			

Q13.29.

1. _____

2. _____

Q13.30. (check one)

_____ I see the strain of the exam is making you stressed.

_____ I see the stress of the exam is straining you.

Q13.31. _____

Q13.32. (circle one)   likely   unlikely

Explanation: _____

_____

Q13.33. (circle one)   yes   no   can't tell

Q13.34. (circle one)   yes   no

Q13.35. (check one)

_____ It probably outcrops on adjoining quadrangles, and where it does, it is beneath the Tuscarora Sandstone.

_____ There is no telling what leads geologists to their decisions.

Q13.36. (check one)

_____ lakes and reservoirs

_____ streams

_____ the slopes of ridges

Q13.37. _____

Q13.38. (check one)

_____ Yes, that makes sense.

_____ No, geologic processes are a phenomenon of the past, never the present.

Q13.39. (check one)

_____ lakes and reservoirs

_____ streams

_____ the slopes of ridges

Q13.40. _____

Q13.41. (check one)

_____ extremely thick compared to rock units

_____ about as thick as an average rock unit

_____ thin, indicating shallow dimensions

Q13.42. (circle one)   normal   reverse or thrust

Q13.43. (circle one)   extensional forces   compressional forces

Q13.44. (circle one)   extensional forces   compressional forces

Q13.45. (circle one)   rift valley   hot spot   collision or convergence

Q13.46. _____

Q13.47.

1. _____

2. _____

3. _____

4. _____

Q13.48. _____ Period.

Q13.49. (check one)

_____ shallow surface mining (strip mining)

_____ mining that starts at the surface but quickly goes deep (underground mining)

Q13.50. (check one)

_____ large strike-slip faults

_____ strong folding of the rocks

_____ impact craters

Unit 13   Rock Deformation and Geologic Maps

Last (Family) Name_____First Name_____

Instructor's Name_____Section_____Date_____

Q13.51. (check one)

_____ They cut across each other.

_____ They parallel each other.

_____ No general relationship exists.

Q13.52. (circle one)   vertical   horizontal   can't tell

Q13.53. _____ feet

Q13.54. about _____ feet

Q13.55. (circle one)   yes   no

Q13.56. (circle one)   yes   no

Q13.57. (circle one)   NE   NW   not given

Q13.58. _____ to _____ degrees to the southeast

Q13.59. _____

The symbol stands for: _____

Q13.60. _____

Q13.61, 13.62, 13.63.

Figure 13.11 **Four incomplete block diagrams.**

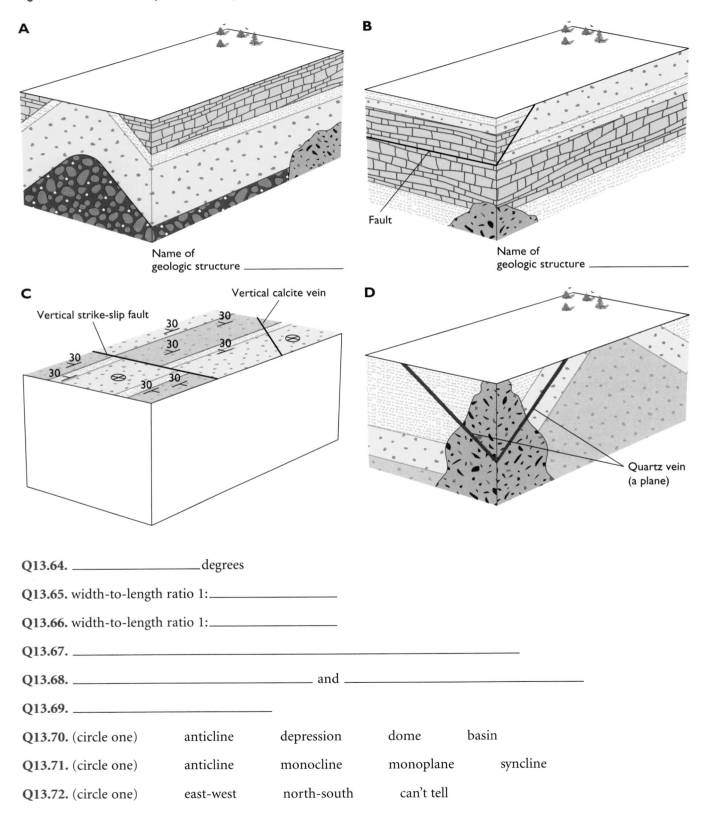

A

Name of
geologic structure _____

B

Fault

Name of
geologic structure _____

C

Vertical strike-slip fault

Vertical calcite vein

D

Quartz vein
(a plane)

Q13.64. _____ degrees

Q13.65. width-to-length ratio 1:_____

Q13.66. width-to-length ratio 1:_____

Q13.67. _____

Q13.68. _____ and _____

Q13.69. _____

Q13.70. (circle one)   anticline   depression   dome   basin

Q13.71. (circle one)   anticline   monocline   monoplane   syncline

Q13.72. (circle one)   east-west   north-south   can't tell

Water: Profoundly Influencing Earth's Surface

Part VI

Water profoundly influences Earth's surface, whether the water is **surface water** in streams or **groundwater** in the rocks and soil underground. The influence of water is both physical and chemical. Physically, water shapes the landscape though erosion, wearing down mountains and carving great canyons. Chemically, water leaches minerals from soil and rocks and transports molecules to other locations.

In Unit 14, you will examine streams and the concepts of drainage basins, stream sediment, and stream flooding. In Unit 15, you will see how water moves underground and how pollution in groundwater can silently move into water wells.

Streams and Flooding Unit 14

Tiny rills, lazy creeks, rushing white-water streams, and huge, muddy rivers—geologists call them all by the general name streams. A **stream** is water running in a channel of any size. Water level in a stream rises and falls, and its watercourse (path) shifts over time.

You need to know the basics of stream behavior, not only as part of your science education, but for personal safety. Large or small, all streams flood when conditions are right. There are gradual floods caused by snowmelt and spring rains, and there are dangerous flash floods. For example, in 1976, the Big Thompson River in Colorado, bloated by sudden, enormous thunderstorms, rapidly rose from its ordinary depth of 18 inches to 32 feet, sweeping more than 700 people to their deaths.

No matter where you live—even in the desert—you probably will experience a major flood in your lifetime.

Lab 14.1 Sorting of Stream Sediments

The size of particles in Earth sediment varies greatly, from boulders as big as a truck to grains too small to see. The distribution of particle size in a sediment may also vary. Both particle size and distribution of sizes hold important information about how a sediment formed.

Materials
- sediment samples labeled **1, 2,** and **3**
- sieves
- mass balance (scale)
- brush, spatula, and/or spoons
- ruler
- magnifying glass or hand lens

Procedure
1. You have been given three samples of sediment from streambeds. Weigh each sample.

Q14.1. Record the weight of each sample in Table 14.3 on your Lab Answer Sheet. Include the units of your measurement and your best estimate of the degree of error.

2. Select the coarsest sieve (biggest holes). Using a ruler, count how many holes (squares) occur over 1 inch of the grid. Then, divide 1 inch by the number of holes to find the size of each hole. (*Example:* If you count 4 holes in 1 inch of the grid material, each is $1/4$ inch in size.)

Q14.2. Record this number in Table 14.3.

3. Measure the holes in the rest of the sieves in your set. For the finer sieves you may use a magnifying glass or information from your instructor.

Q14.3. Record the sizes of the holes in the other sieves in Table 14.3.

4. Set the sieves one on top of the other, with the coarsest sieve on top, the next finer below it, and so on down to the bottom collecting pan.

5. Carefully sieve sample **1**. Remember that your patience will accomplish what force cannot: gently tap, tilt, and jiggle your sieves from side to side.

6. When sample **1** is fully sieved, separate the sieves. Weigh the sediment in each sieve.

Q14.4. Record the weights in Table 14.3.

Q14.5. In the table, add all the measurements for sample **1** ("Arithmetic total"). (*Note:* If your total for any sample differs by more than a few percent from your original sample weight, find your error. Consult your instructor. If necessary, reweigh the separated samples. You also may need to reweigh the whole sample of the sediment fractions combined.)

7. When your work is within what your instructor says is acceptable error, repeat the entire procedure for samples **2** and **3**.

8. Convert the weight of each size-fraction of sample **1** into a percentage.

Q14.6. Write each percentage in Table 14.3.

9. Repeat for samples **2** and **3**.

Q14.7. Using the percentages you just calculated, plot histograms of the particle size distribution for each sample. Use the grids provided in Figure 14.4 on your Lab Answer Sheet.

Applying What You Just Experienced

Q14.8. Looking at the histograms you produced, which of your samples has the greatest percentage of its weight composed by a single size fraction?

Q14.9. All things considered, which sample do you think is best sorted? Explain.

Q14.10. Which, if any, of your samples show a bimodal distribution?

Q14.11. What geologic environment(s) might best account for what you have documented about particle size and particle size distribution for each of your samples?

Lab 14.2 Measuring Stream Velocity and Discharge

The velocity and discharge of a stream can be crucial information for anyone living near it or boating on it. This exercise gives you real-world experience in measuring the key variables that control stream behavior. (*Note:* Keep safety in mind throughout this lab! You must look out for yourself and your classmates first. Collecting data is a secondary concern.)

Materials
- hammer for driving stakes
- yardstick or ruler
- stakes
- stopwatch or watch with second hand
- string
- permanent marking pen
- optional: propeller-type velocity measuring instrument
- optional: waders

Procedure
1. Visit a nearby stream to measure the data needed to calculate the stream's discharge. You will be divided into teams and given a place at which to do this exercise.

2. Drive your two stakes at the water's edge on opposite sides of the stream. Make sure the line between the two stakes is perpendicular (at a right angle) to the direction of stream flow.

3. Using a permanent marker, mark the string at 1-foot intervals. Make several more marks than the width of the stream. (*Example:* If the stream is about 10 feet wide, make 13 or 14 marks on the string.)

4. Tie the string to one stake so that one of the first marks is just at the stake.

5. Stretch the string across the stream and tie it to the second stake. (***Safety note:*** Do not wade into water deeper than knee level at any point!)

Q14.12. Record the width of the stream in Table 14.4 on your Lab Answer Sheet (*Note:* If the second stake is not at a foot mark, estimate the final fraction of a foot).

6. Think of your first stake as your starting point, your "0 distance" reference point. Note that the water depth at this stake is zero. These data have already been recorded in Table 14.4.

7. Move to the first 1-foot mark on the string. Using a yardstick, measure the water depth from the water's surface (immediately below the mark on the string) to the stream bottom. Use feet as your unit.

Q14.13. Call out your results for a teammate to record in Table 14.4.

8. Continue across the stream, repeating step 7 at each foot mark. (***Safety note:*** If the water seems too deep or too swift, STOP and consult your instructor.)

9. When you reach the second stake, tell your instructor that you have finished the depth measurements. (Don't worry if you have not used all the blanks in the table.)

10. Now measure the velocity of the stream at each of your foot marks. Use either the instrument provided or an approximation method (explained by your instructor).

Q14.14. Record your results in the table in feet per second.

11. Share and verify your data with your teammates.

Applying What You Just Learned
Q14.15. Refer to Figure 14.5 on your Lab Answer Sheet. Mark the spot where your second stake was placed. (It should be on the same horizontal line as the first stake, which is already marked on the figure.) Next, mark dots to indicate the depth of the water at each 1-foot interval you measured. (***Note:*** Round your data to the nearest quarter-foot.) Draw a line to connect your depth dots. This creates a cross section of the stream.

Each box on the grid represents 1 square foot. Count all the full boxes in your stream's cross section and note your result. Add any partial boxes, rounding to the nearest quarter-foot.

Q14.16. What is the total area of your stream's cross section?

Q14.17. What is the mean (average) of your stream's velocity? (***Hint:*** Consider the significance of the zero velocity at both stakes.)

Q14.18. From your data, what is the stream's discharge at this point?

Q14.19. During what times of the year or under what conditions do you think your stream might have much greater discharge at the same place?

Q14.20. What part or parts of the stream as you measured it currently have the greater velocities?

Problems

Problem 14.1. Drawing Some Drainage Basins of South America. Figure 14.6 on your Problem Answer Sheet shows many of the major streams in South America.

Q14.21. Using a red pencil, outline the **drainage basin** of the Amazon River. Lightly shade the area within the basin.

Q14.22. Approximately what proportion of South America is within the Amazon drainage basin?

Q14.23. Using a green pencil, outline and shade the drainage basin of the Rio de la Plata.

Q14.24. Using a blue pencil, heavily mark the line that divides the two drainage basins you outlined. This line is called a **drainage divide.**

Q14.25. The longest single stream channel in a drainage basin is the **trunk stream.** Using your blue pencil, mark the trunk stream of the Amazon River basin.

Problem 14.2. Drawing a Stream Cross Section. Some geology students measure a stream's average velocity as 0.4 foot per second. They also measure the stream's depth at 1-foot intervals on a line directly across the stream. Their measurements follow.

At edge of bank on one side	0 feet (no water)
1 foot out over stream	$1/2$ foot of water depth
2 feet out over stream	1 foot of water depth
3 feet out over stream	1 foot of water depth
4 feet out over stream	$1/2$ foot of water depth
5 feet out over stream	$1/2$ foot of water depth
At edge of bank on other side	0 feet (no water)

Q14.26. Make a scale drawing of the stream cross section using the grid in Figure 14.7 on your Problem Answer Sheet.

Q14.27. How many total blocks are included in the area of your stream cross section? (Each block is a square foot, so your answer will be in square feet.)

Q14.28. Using your answer, calculate the discharge of the stream.

Problem 14.3. Sediment Size and Stream Velocity. At the edge of a stream, you scoop a handful of sand and examine it with a hand lens. You estimate the sand grains to be about 1 millimeter in diameter.

Q14.29. From the information in Figure 14.1, what is the minimum stream velocity required to begin transporting 1-millimeter-diameter sand grains? (***Note:*** Answer in meters per second.)

Q14.30. The gravel along another stream's edge is about 1 centimeter in diameter. What minimum stream velocity would you expect to start transporting 1-centimeter-diameter gravel? (***Note:*** Answer in meters per second.)

Q14.31. Convert your answers to the two preceding questions to feet per second.

Problem 14.4. Proportions of Water in a Drainage Basin. Cougar Creek's drainage basin totals 100 square miles. The area's average rainfall is 24 inches per year. At its mouth, the creek's flow averages 100 cubic feet per second.

Q14.32. If the creek's flow is composed of annual run-off and infiltration, what proportion of the annual rainfall is lost to evaporation and evapotranspiration?

Problem 14.5. Stream Flooding. A stream flowing through a midwestern town has a gauging station. At this station, the U.S. Geological Survey recorded stream flow from the 1930s until funding ran out in the 1970s. Table 14.1 shows data for the stream's peak annual discharge.

Q14.33. In what year did the stream experience its greatest peak discharge?

Q14.34. In what year did the stream experience its smallest peak discharge?

Q14.35. How many times larger was the largest annual discharge compared to the smallest?

Geologists can determine the **recurrence interval** of flooding for a stream. The recurrence interval indicates how often a flood of a specific size occurs, based on historical records. (See your textbook for details.) Geologists use this equation:

$$\text{recurrence interval} = \frac{N + 1}{M}$$

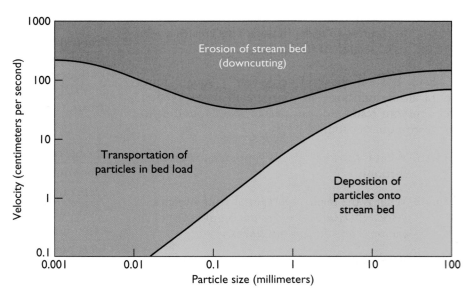

Figure 14.1 **Stream velocity versus particle size and movement.**

Table 14.1

Peak annual discharge (cubic feet per second)

1938	972	1948	927	1958	782	1968	880
1939	911	1949	682	1959	888	1969	912
1940	889	1950	812	1960	914	1970	984
1941	1033	1951	948	1961	707	1971	976
1942	941	1952	991	1962	1308	1972	840
1943	770	1953	776	1963	1112	1973	691
1944	1211	1954	882	1964	964	1974	922
1945	946	1955	756	1965	729		
1946	792	1956	912	1966	741		
1947	841	1957	1104	1967	972		

where N = number of years of stream discharge records and M = rank-order magnitude for a specific flood (1 = greatest in the record, 2 = next greatest, and so on.)

Q14.36. What is the recurrence interval of the flood of the largest peak annual discharge?

Q14.37. What is the recurrence interval of the 1941 flood?

Q14.38. What is the recurrence interval of the 1949 flood?

Q14.39. If records were kept from 1974 to the present day, would recurrence interval calculations have more meaning or less meaning? Why?

Problem 14.6. Snake River Flooding. The Snake River is one of the largest in the Pacific Northwest, draining most of Idaho and southeastern Washington State. When explorers Merriwether Lewis and William Clark reached the Snake River on their westward journey, they built dugout canoes and floated down the Snake to the Columbia River and on to the Pacific Ocean.

Where the Snake River crosses from Idaho into Washington, its drainage basin measures 103,200 square miles. For 44 years, from 1929 to 1972, the U.S. Geological Survey recorded the Snake's peak annual discharge at this point (Table 14.2 on the next page).

Q14.40. Graph the annual peak discharge data in Figure 14.8 on your Problem Answer Sheet.

Q14.41. In what year did the Snake River (at this point) reach its greatest maximum annual discharge?

Q14.42. In what year was the maximum annual discharge the smallest?

Q14.43. How many times smaller was the smallest peak annual discharge than the largest?

Geologists characterize a major flood as a "50-year flood" or a "100-year flood" (see your textbook). The idea is that, from historical records like those in Table 14.2, we can determine the likelihood of a flood of a particular severity over time. As noted in problem 14.5, geologists use the formula

$$\text{recurrence interval} = \frac{N + 1}{M}$$

where

N = number of years of stream discharge records
M = rank-order magnitude for the flood in question.

Q14.44. Refer to Table 14.2 (on the next page). Using the recurrence interval formula, what is the recurrence interval for the flood of 1948?

Table 14.2

Peak annual discharge of the Snake River at the Idaho-Washington border, 1929–1972 (cubic feet per second)

1929	155,000	1940	126,000	1951	182,000	1962	140,000
1930	95,600	1941	102,000	1952	250,000	1963	154,000
1931	116,000	1942	162,000	1953	232,000	1964	266,400
1932	219,000	1943	209,000	1954	210,000	1965	247,000
1933	245,000	1944	109,600	1955	204,000	1966	114,000
1934	164,000	1945	149,000	1956	292,000	1967	215,000
1935	130,000	1946	169,000	1957	323,000	1968	134,000
1936	219,000	1947	239,000	1958	248,000	1969	189,000
1937	114,000	1948	369,000	1959	171,000	1970	233,000
1938	219,000	1949	248,000	1960	164,000	1971	258,000
1939	149,000	1950	212,000	1961	174,000	1972	240,000

Q14.45. What is the recurrence interval for the flood of 1957?

Q14.46. What is the recurrence interval for a flood the size of the 1930 flood? (**Hint:** You can find M by eyeballing the list of data.)

Q14.47. On the East Coast, some stream flooding records extend much farther back into history than in the West. Does this mean that recurrence intervals would have more or less meaning in the East? Why?

Q14.48. Recent decades have seen building booms in urban areas of North America. Wetlands have been filled so that homes, streets, and shopping malls could be constructed. How do such changes affect flood frequency?

Problem 14.7. Topography and Streams in a Dry Climate. Study the topographic map of southern California in Figure 14.2. Note the intermittent streams. To understand the dangers of flooding in this area, you first must consider the topography.

Q14.49. What is the highest elevation shown on the map? (**Hint:** The highest point on a map may not be the top of a hill.)

Q14.50. What is the lowest elevation shown in Reche Canyon?

Q14.51. What is the total relief of the area?

Q14.52. Imagine that your friend rents a unit in the trailer park shown on the map. You say, "This place is just asking for a flood!" But your friend answers, "No, the trailer park manager told me that the park has never flooded. Look, the stream is actually dried up!" Who has the better understanding of the implications of the topography of the area?

Q14.53. What evidence indicates modest efforts at flood control in Reche Canyon?

Q14.54. If the trailer park floods, is it likely to be a gradually rising flood over several days or a flash flood?

Q14.55. Imagine that an enormous thunderstorm inundates most of Section 11 on the map (the red square labeled "11"). As the storm ends, you and your bicycle are southeast of the trailer park at the intersection of the main road through Reche Canyon and the short road that runs eastward toward the edge of Section 2. Knowing what you do about flooding, which way should you bike to save your life?

Problem 14.8. Recurrence Intervals and Discharge in Nashville, Tennessee. Discharge data from the Cumberland River in Tennessee are plotted in Figure 14.3. A government agency wants to build a spillway for the

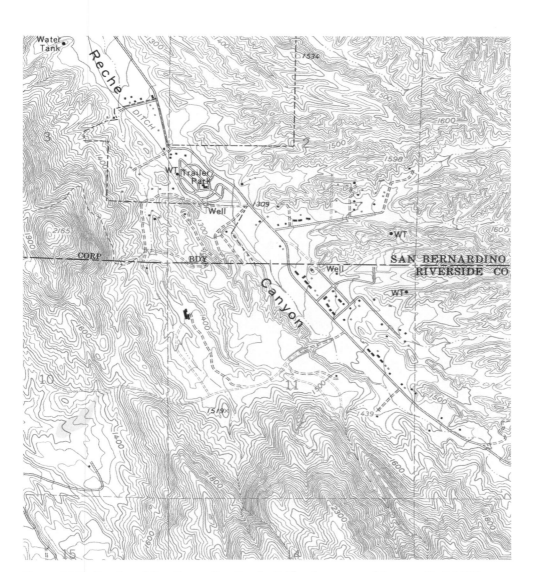

Figure 14.2 Portion of San Bernardino South, California topographic map. *(1 : 24,000 scale, revised 1980. Courtesy of U.S. Geological Survey)*

river that can contain a flood up to and including a 200-year flood.

Q14.56. Using a ruler to make a straight-line extrapolation of the data in the figure, what is the approximate discharge capacity the spillway should be designed to carry?

Q14.57. If you assume that the data are better fitted to a curve than to a straight line, would your estimate be different?

Q14.58. Even using a good historical record, is it better to regard a discharge projection as a guarantee or as a reasonable estimate?

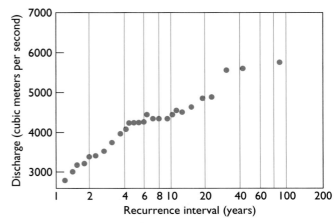

Figure 14.3 Recurrence intervals and discharge for Cumberland River in Tennessee.

Unit 14   Streams and Flooding

Last (Family) Name_____ First Name_____

Instructor's Name_____ Section_____Date_____

Q14.1, 14.2, 14.3, 14.4, 14.5, 14.6

Table 14.3

Sorting stream sediments and weighing samples

Size of holes (fraction of an inch)	Sediment sample 1 starting weight _____	Sediment sample 2 starting weight _____	Sediment sample 3 starting weight _____
Coarsest sieve: _____ inch	_____ wt _____ wt % of whole sample	_____ wt _____ wt % of whole sample	_____ wt _____ wt % of whole sample
Second-coarsest sieve: _____ inch	_____ wt _____ wt % of whole sample	_____ wt _____ wt % of whole sample	_____ wt _____ wt % of whole sample
Third-coarsest sieve: _____ inch	_____ wt _____ wt % of whole sample	_____ wt _____ wt % of whole sample	_____ wt _____ wt % of whole sample
Fourth-coarsest sieve: _____ inch	_____ wt _____ wt % of whole sample	_____ wt _____ wt % of whole sample	_____ wt _____ wt % of whole sample
Bottom collecting pan	_____ wt _____ wt % of whole sample	_____ wt _____ wt % of whole sample	_____ wt _____ wt % of whole sample
Arithmetic total of weights, all size fractions			
Difference between arithmetic total and original weight, converted to %			

Q14.7.

Figure 14.4 Histograms of results.

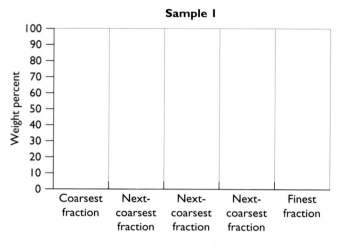

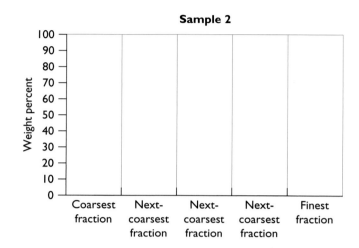

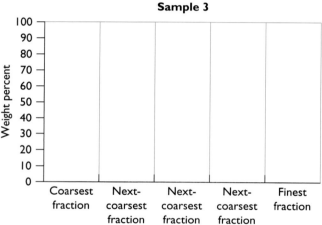

Q14.8. (circle one)   sample **1**   sample **2**   sample **3**

Q14.9. (circle one)   sample **1**   sample **2**   sample **3**

Explanation: _____

_____

Q14.10. (circle one)   sample **1**   sample **2**   sample **3**   none

Q14.11. Circle the sample number(s) for each geologic environment.

Sample			Environment
1	2	3	Fast-moving, white-water stream in mountainous terrain
1	2	3	Medium-to-large stream with lower gradient
1	2	3	Large, mature stream meandering down very low gradient conditions

Q14.12, 14.13, 14.14.

Table 14.4

Stream data from field work | | **Width of stream = _____ feet**

Distance from first stake (feet)	Depth (feet)	Stream velocity (feet per second)
0	0	0
1		
2		
3		
4		
5		
6		
7		
8		
9		
10		
11		
12		
13		
14		
15		
16		
17		
18		
19		
20		
21		
22		
23		
24		
25		
26		
27		
28		
29		
30		

Q14.15.

Figure 14.5 **Cross section of stream.**

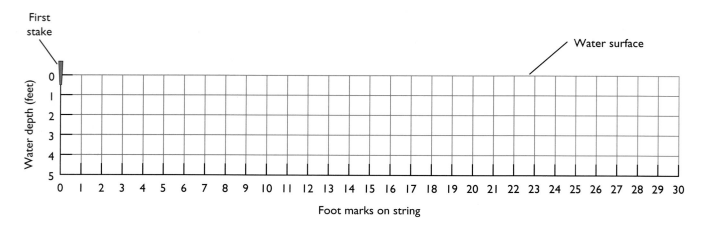

Q14.16. _____ square feet

Q14.17. _____ feet per second

Q14.18. _____ cubic feet per second

Q14.19. _____

Q14.20. _____

_____

Unit 14   Streams and Flooding

Last (Family) Name_____ First Name_____

Instructor's Name_____ Section_____ Date_____

Q14.21, 14.23, 14.24, 14.25. (Q14.22 is on the following page.)

Figure 14.6   **Major streams of South America.**

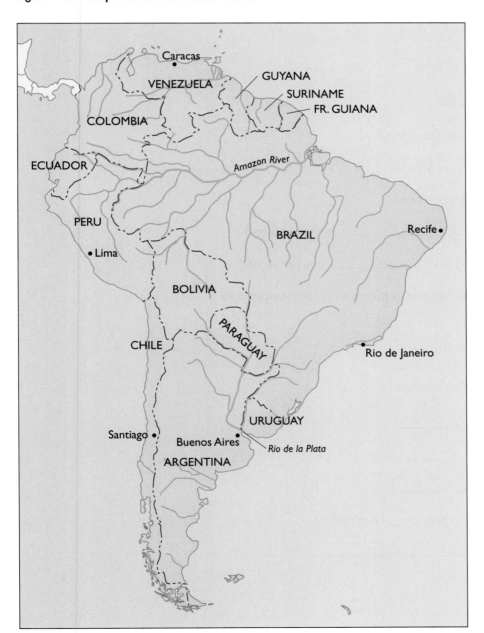

Q14.22. (circle one)   0.25%   2.5%   25%   99%

Q14.26.

Figure 14.7 Scale drawing of stream cross section.

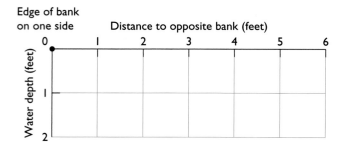

Q14.27. _____ square feet

Q14.28. _____ cubic feet per second

Q14.29. _____ meters per second

Q14.30. _____ meters per second

Q14.31. 1-millimeter-diameter sand: _____ feet per second

1-centimeter-diameter gravel: _____ feet per second

Q14.32. _____ % is lost to evaporation and evapotranspiration

Q14.33. _____

Q14.34. _____

Q14.35. _____ times

Q14.36. _____

Q14.37. _____

Q14.38. _____

Q14.39. (circle one)   more   less   can't tell

Explanation: _____

Q14.40.

Figure 14.8   **Plot of discharge of Snake River at the Idaho-Washington border**.

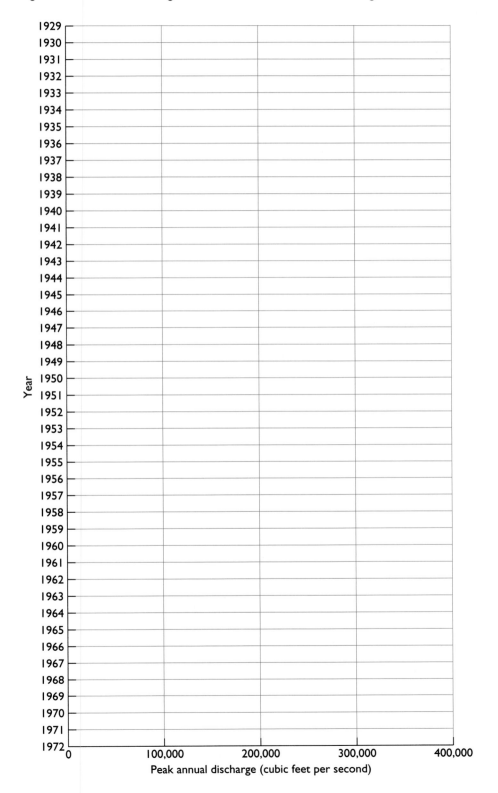

Q14.41. _____

Q14.42. _____

Q14.43. _____ times smaller

Q14.44. Recurrence interval for 1948 flood = _____ years

Q14.45. Recurrence interval for 1957 flood = _____ years

Q14.46. Recurrence interval for 1930 flood = _____ years

Q14.47. (circle one)   more   less   can't tell

Explanation:_____

Q14.48. _____

_____

Q14.49. _____ feet

Q14.50. _____ feet

Q14.51. _____ feet

Q14.52. (circle one)   you   manager   can't tell

Q14.53. _____

Q14.54. (circle one)   gradual   flash

Q14.55. (check one)

_____ downhill (NW) on the main road

_____ uphill (SE) on the main road

_____ uphill (ENE) on the short road

Q14.56. _____ cubic meters per second

Q14.57. (check one)

_____ yes, possibly quite different

_____ no

Q14.58. (circle one)   guarantee   reasonable estimate

Groundwater is underground, so it is out of our sight and can easily slip out of our thoughts. But in some areas of North America, groundwater actually determines the shape of the land (topography). A dramatic example is when groundwater creates a **sinkhole.** Large sinkholes (most notably in Florida) have swallowed houses or cars—even an entire automobile dealership!

   Less dramatic aspects of groundwater geology are economically crucial. For example, in arid climates across North America, groundwater is the main source—sometimes, the only source—of drinking water for major cities and crop irrigation. Consequently, even tiny concentrations of pollutants within groundwater cause concern.

Lab 15.1   Sediment Porosity Depends on Particle Shape

Porosity depends partly on particle shape. In this experiment, you will use particles of different shapes but about the same size. You will also consider how sediment sorting can influence porosity.

Materials
- sediment sample of rounded pebbles
- sediment sample of angular gravel
- graduated beakers or cylinders, two sizes
- water

Procedure
1. Put your rounded pebbles into the larger graduated container. Gently tap it to settle the pebbles.

2. Measure the volume occupied by the pebbles by visually aligning the top surface of the pebbles with a graduation mark on the container.

Q15.1. Record your answer in Table 15.1. on the Lab Answer Sheet.

3. Fill the smaller graduated container with water and note the volume of water here: _____. Carefully pour the water over the pebbles, filling the container until the water just covers all the pebbles. (***Note:*** Measure carefully and use the meniscus of the

water for your measurement. If you add too much water, pour it out and repeat this step.)

4. Note the volume of water remaining in the smaller cylinder: _____. Calculate how many milliliters of water you added to fill all the pores around the pebbles.

Q15.2. Write your answer in Table 15.1.

5. Calculate the porosity of the pebbles. The formula

$$\frac{\text{total volume of pores (the volume of water you added)}}{\text{volume of the material (the volume of the dry sphere)}} \times 100 = \% \text{ porosity}$$

for porosity is:

Q15.3. Record your answer in Table 15.1.

6. Repeat steps 1–5 for the angular gravel.

Q15.4. Enter your results for the angular gravel in the table.

Applying What You Just Experienced
This experiment used particles of generally the same size but different shapes. You can see that when particle *size* is constant, particle *shape* is important for determining sediment porosity.

Q15.5. Which of your samples had the greater porosity, the rounded pebbles or the angular gravel?

Q15.6. Imagine that you are living in an arid climate. You are going to build a house in the country, and you know you will have to drill a well for your water supply. For your well to be successful, you must drill into sediment or rock that is porous so that it can hold the water you need. Which type of sediment near your house would likely hold more water, a deep layer of well-rounded pebbles or a deep layer of angular gravel?

   Just as the porosity of sediment depends on particle shape, it also depends on *sorting.* Poorly sorted sediment (sediment that is mixed in size) has less porosity because it includes smaller particles that fill the pores around larger particles. Well-sorted sediment has greater porosity. Indeed, sorting is often the single most important variable for determining porosity in sediment.

For a better understanding of this concept, see Figure 15.4 on your Lab Answer Sheet. It shows two beakers containing marble-sized sediment. In beaker 2, mark numerous dots to represent sand grains, *doing so only in the empty "pore" spaces between the larger particles.* When you do this, you are "filling in" the pore spaces. Complete your work with dots in beaker 2.

Q15.7. Which beaker now contains the better-sorted sediment?

Q15.8. Which beaker contains the more porous sediment and therefore can hold more water?

Lab 15.2 Capillary Action— in Action!

Your instructor has set up a glass device that clearly shows the effect of capillary forces. Note the different heights the water attains in the different-size tubes.

Weak capillary forces attract water molecules to the surface of solids. This draws water up the tubes. It also creates a "capillary fringe" of water in soil and rock pores above the water table.

Procedure
1. Measure the inside diameter of each of the tubes in millimeters.

Q15.9. Record your answer in Table 15.2 on the Lab Answer Sheet.

2. Measure the height of the water (in millimeters) in the tubes compared to the water surface in the largest tube (which we'll assign a value of 0).

Q15.10. Record your data in Table 15.2.

Applying What You Just Experienced
The water table, when viewed up close, shows the effect of capillary action. Fine-grained materials like sand and soil hold water just above the water table due to capillary forces. Sand and soil have many tiny pores that we can think of as tiny "tubes." If we were the size of amoebas, we would doubtless pay much more attention to capillary action! But we are very large relative to capillary forces and thus sometimes neglect their importance.

Q15.11. Looking at your data, would you say that capillary forces increase in a linear fashion as the diameter of a passageway decreases? Or would you say the relationship is some sort of exponential one? (**Hint:** You may wish to plot your data on scrap paper to evaluate it.)

Q15.12. Where water is held in the capillary fringe above a water table, would it flow freely into a water well?

Problems

Problem 15.1. Hydraulic Gradient. Figure 15.1 is a schematic diagram showing two wells. Well **A** was drilled at a ground elevation of 1200 meters. The well intersects the water table at a depth of 15 meters. The wellhead for well **B** rests at 950 meters. Well **B** intersects the water table at a depth of 5 meters.

Q15.13. Consulting your notes or textbook, calculate the hydraulic gradient between the wells. Express your answer in meters per kilometer.

Problem 15.2. Rates of Groundwater Movement. You can roughly measure rates of groundwater movement by injecting a dye into groundwater at one well. Then you observe the water in a nearby well. When the dye appears in the water of the nearby well, you note the time. This tells you how long it takes groundwater to flow between the wells. If you also measure the distance between the wells, you can determine the **flow rate** in

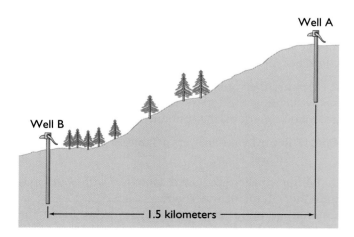

Figure 15.1 Hydraulic gradient.

that direction: so many inches per day or feet per week or however you need to express it.

In gravel beds near the surface, groundwater moves relatively fast. A typical velocity might be 25 centimeters per day (10 inches per day).

Q15.14. Imagine that you live in the country and your neighbor's septic system is 0.25 kilometer up the groundwater flow path from your water well. If the groundwater is in gravel beds and the water flows at a typical rate, how long would it take for tasty dissolved material from your neighbor's septic system to reach your water well?

In solid rock, groundwater generally travels at much slower rates. In some granites, for example, movement may be as little as 0.05 centimeter per day.

Q15.15. If you and your neighbor lived in houses built on such granite, how long would it take for dissolved matter to reach your water well?

Q15.16. Convert your answer into years.

Problem 15.3. Toxic Dumps and Groundwater. Study Figure 15.2. and assume a flow path from right to left in the figure.

Q15.17. Can toxins from the small toxic waste dump reach the water well?

Q15.18. Could dissolved matter and bacteria from the septic system ever get into the well? Explain. (***Hint:*** Consider cones of depression around the well.)

Problem 15.4. Darcy's Law. Darcy's law states:

$$\frac{Q}{A} = \frac{K \times h}{l}$$

where

Q = volume of water flowing in a certain time
A = cross-sectional area through which the water is flowing
h = vertical drop in the water table
l = horizontal flow distance between two points

Q15.19. If h is expressed in meters, A is expressed in square meters, l is expressed in kilometers, and Q is expressed in cubic meters per hour, what are the units for K?

Q15.20. What is K called?

Problem 15.5. Topographic Map of Mammoth Cave, Kentucky. Study the topographic map in Figure 15.3 on page 229.

Q15.21. How many places on the Mammoth Cave map are labeled "entrances" to the cave?

Q15.22. The map scale is 1:24,000. Assume that the cave is at least as long as the distance from the Historic Entrance to the New Entrance. What is the minimum length of the cave in kilometers?

Q15.23. Referring to your textbook, from what type of rock is this vast cavern most likely formed?

Q15.24. What process is likely to have removed such large quantities of rock to form the cavern?

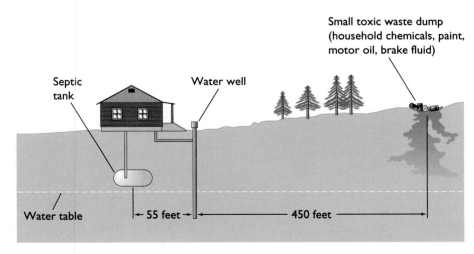

Figure 15.2 Common real-life groundwater pollution problem.

Q15.25. What do we call features like Double Cellars? How do such features form?

Q15.26. Flint Ridge is in the upper right corner of the map. What is the highest topographic contour line on Flint Ridge? (If it is not directly numbered, figure out its elevation.)

Q15.27. What is the elevation of the benchmark next to the Green River, just west of the Historic Entrance?

Q15.28. What is the difference in elevation between the highest contour on Flint Ridge and the benchmark along the big river?

Q15.29. Considering the significant relief you just calculated and the fact that Kentucky is in a humid part of the country, would you expect permanent, swiftly flowing streams in this area?

Q15.30. What is, in fact, happening to almost all surface water near Flint Ridge?

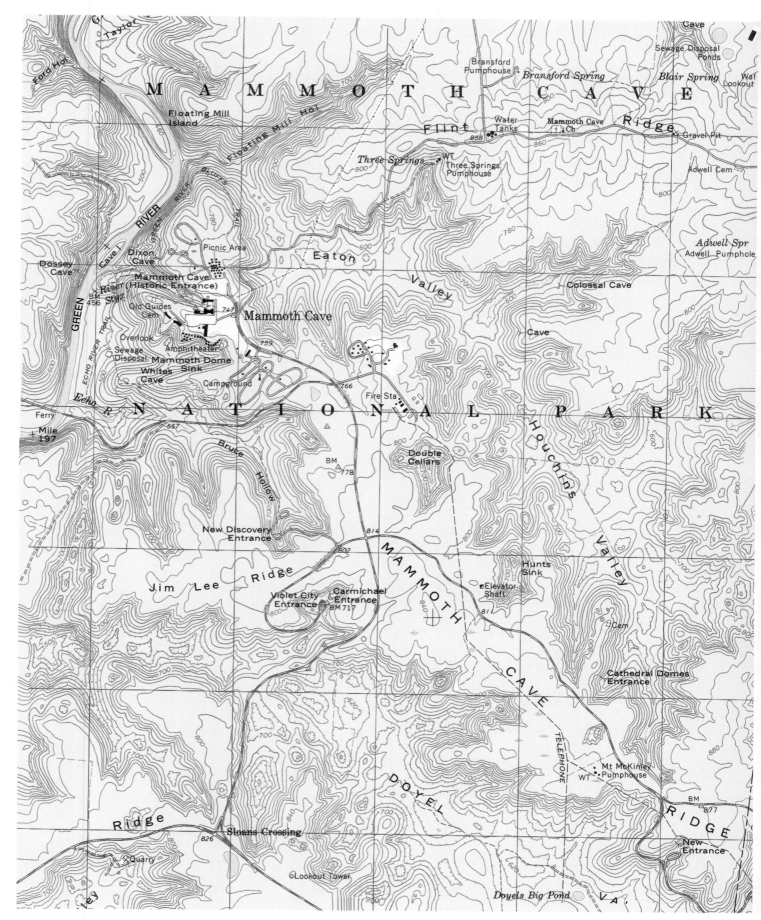

Figure 15.3 **Topographic map of Mammoth Cave, Kentucky.** *(1:24,000 quadrangle. Courtesy of U.S. Geological Survey).*

Unit 15 Groundwater

Last (Family) Name_____ First Name_____

Instructor's Name_____ Section_____ Date_____

Q15.1, 15.2, 15.3, 15.4.

Table 15.1

Porosity of geologic materials

	Volume of material (milliliters)	Volume of water added (milliliters)	Sample porosity (percent)
Rounded pebbles			
Angular gravel			

Q15.5. (check one)

_____ rounded pebbles

_____ angular gravel

Q15.6. (check one)

_____ rounded pebbles

_____ angular gravel

Figure 15.4 Particle sorting affects porosity.

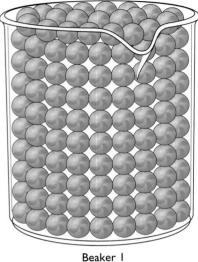 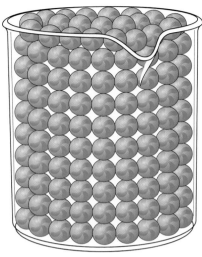

Beaker 1   Beaker 2

Q15.7. (circle one)   beaker 1   beaker 2

Q15.8. (circle one)   beaker 1   beaker 2

Q15.9, 15.10.

Table 15.2

Height of water in tubes

	Largest tube	**Next-smaller tube**	**Next-smaller tube**	**Smallest tube**
Inside diameter of tube (millimeters)				
Height of water compared to height in largest tube (millimeters)	0			

Q15.11. (circle one)   linear   exponential

Q15.12. (check one)

_____ yes, because of gravity

_____ no, because the water is attracted to the surfaces of solids around it

Unit 15 Groundwater

Last (Family) Name_____ First Name_____

Instructor's Name_____ Section_____Date_____

Q15.13. _____ meters/kilometers

Q15.14. _____ days

Q15.15. _____ days

Q15.16. _____ years

Q15.17. (circle one)   yes   no

Q15.18. (circle one)   yes   no

Explanation: _____

_____

_____

Q15.19. _____

Q15.20. K is the _____.

Q15.21. _____ entrances.

Q15.22. _____ kilometers

Q15.23. _____

Q15.24. _____

Q15.25. They are called_____. They form due to _____

_____.

Q15.26. _____ feet

Q15.27. _____ feet

Q15.28. _____ feet

Q15.29. (check one)

_____ Yes, this much relief in a wet climate would generally yield a lot of significant streams.

_____ No, I would never expect streams in a wet climate like this one.

Q15.30. _____

Geologists now recognize two general rates for Earth processes: gradual and catastrophic (fast). Some processes fall in between, but the two extremes are a useful way to think about the rates at which Earth changes. **Gradual processes** occur over time spans too great to observe in a lifetime. Gradual changes include the weathering of rocks and the slow erosion of deepening stream valleys. **Catastrophic processes,** on the other hand, can be observed on a human time-scale. They include volcanic eruptions and meteorite impacts.

Catastrophic Change: Impact Craters and Outburst Flooding **Unit 16**

Catastrophic processes occur fast enough that we humans can see them play out. They occur over relatively short time spans, from seconds to months. Examples are landslides, earthquakes, and meteorite impacts.

Both gradual and catastrophic processes can cause great change—only the *rate of change* differs. In this unit, you will experiment with simulations of two events that promote catastrophic geological change: meteorite impacts and glacial outburst flooding. (In Unit 17, "Gradual Change," we'll look at gradual processes of change, such as erosion by glaciers, wind, and ocean waves, and changes in climate over geologic time.)

Lab 16.1 Transformation of Kinetic Energy into Thermal Energy

When a meteorite strikes Earth, its flight through space comes to an abrupt stop. All of its **kinetic energy** (energy of motion) is converted almost instantly into **thermal energy** (heat). The amount of energy released by the impact of a large meteorite is staggering—much greater than the energy of many nuclear bombs combined.

Materials
- hammer
- broad-headed nails
- small-headed finishing nails
- hardwood board or particle board

Procedure

✔ SAFETY CHECK ▸ It's not just your fingers that are at risk when using a hammer. More serious is the possibility of eye injury from flying bits of wood or metal from the hammer's impact. Wear eye protection. And please use the hammer with caution!

1. Rapidly drive a broad-headed nail fully into the wood.

2. As soon as the nail is fully sunk into the wood, quickly touch the head of the nail and the face of the hammer, sensing the temperature of each.

Q16.1. Record your observations in Table 16.1 on your Lab Answer Sheet.

3. Repeat the procedure with a small-headed finishing nail.

Q16.2. Record your observations in Table 16.1.

Applying What You Just Experienced
Q16.3. What is the name for the type of energy represented by the swinging hammer?

Q16.4. In the nail head, the energy of the hammer blows has been transformed into what other kind of energy?

Q16.5. Which became warmer, the broad-headed nail or the small-headed nail? Why?

Q16.6. Why did the temperature of the hammer rise less than the temperature of the nails?

Q16.7. In general, meteorites strike Earth while moving 5 to 15 kilometers per second. Is this faster or slower than your hammer moves as you swing it?

Q16.8. If a meteorite with the same mass as your hammer hit Earth, would more or less heat be generated than by your hammer blows?

Lab 16.2 Simulation of Sediment Distribution after Outburst Flooding

In eastern Washington State, near Idaho, there lies a remarkable and unique landscape. Deep channels are cut into barren Earth, giving the area the less-than-appealing name Channeled Scablands. The long channels cut down through the wind-deposited soil (loess) and deep into the black basalt beneath. The local name for such a channel is *coulee*—French for "flow of a torrent"—which is appropriate, as you will see.

Viewed from a satellite, the coulees form a braided pattern. We can see the coulees from space because the black basalt rock contrasts with the brown loess. You may know of Grand Coulee Dam, which has created the sixth largest reservoir in the United States. The dam is built near one huge coulee, appropriately named Grand Coulee.

The coulees and other unusual features of the Scablands were formed during the most recent ice age when an ice dam ruptured in western Montana. The dam had held back glacial Lake Missoula. When the dam broke, an enormous torrent of water, mixed with boulders and glacial ice, rushed across what is today northern Idaho. It then cut across the Scablands toward the Columbia River and Snake River (Figure 16.1). Evidence for this enormous **outburst flood** includes the following features.

1. *The coulee channels form a crisscrossing complex.* The Scablands look like a gigantic braided stream, a channel pattern that indicates rapid inundation by water followed by its rapid retreat. Furthermore, many coulees have no streams running down them, and some are extraordinarily flat, without the hint of the usual stream-carved V shape. In fact, their shape is why they are called *coulee* or *channel* rather than *valley*. There are no "normal" valleys throughout the Scablands.

2. *Deep gravel bars exist in the middle of coulees and at the perimeter of the area.* Some gravel bars are around a hundred feet high and could have been formed only by floodwaters much deeper than that. One bar, about ten miles long, has well-sorted basalt gravels that make it clear that the bar is not a glacial moraine. It looks like a waterborne deposit, but it is more than a hundred miles from the Columbia River, the nearest large stream.

3. *The area has cataract cliffs and plunge pools hundreds of feet in diameter.* These now are completely dry or contain only small amounts of water, which seem unable to have formed such massive features. Two examples are Dry Falls, near Grand Coulee, and Palouse River Falls.

4. *Thick strata of fine silts exist upstream in the Columbia*

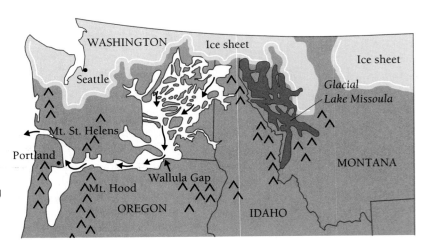

Figure 16.1 Area affected by outburst flooding of glacial Lake Missoula. *(E. K. Peters,* No Stone Unturned. *New York: W. H. Freeman, 1996, Figure 6-2.)*

and Snake Rivers and their tributaries. These tributaries must have flowed *backward* when reached by a wall of outburst floodwater hundreds of feet high. This back-flooding produced the fine sediments preserved in tributary valleys. Beds of fine silt with isolated cobbles can be found up the Snake River Canyon as far away as Idaho.

5. *Giant ridges, up to 50 feet high, exist in the gravel bars in the coulees.* Interpreting these ridges as normal ripple marks on a gigantic scale, geologists can deduce that the coulees must have filled with water hundreds of feet deep, moving in excess of 50 miles per hour.

6. *Longitudinal bars cut across hanging valleys.* These bars were laid down at the edge of the floodwaters, sometimes forming natural levees and cutting across the mouths of what later became hanging valleys. Lakes formed in these valleys after the flood subsided because normal drainage was no longer possible.

7. *Chaotic or deranged drainage patterns exist over much of the area.* The Scablands lack normal drainage and are dominated by closed depressions and coulees. This indicates that the currently active small streams developed later and thus did not form the topography.

This lab and some of the problems will give you a sense of the unusual geologic features of the Channeled Scablands.

Materials
- silts and clays
- fine sand
- coarse sand
- pebbles
- jar with lid
- water

Procedure
1. Mix the sediment together—pebbles, coarse sand, fine sand, silt, and clay. Add the mixture to the jar until it is half full.

2. Add water to almost completely fill the jar. Put on the lid.

3. Thoroughly mix the contents (sediment and water) by rapidly inverting the jar many times.

4. Set the jar on a table and observe it closely.

Q16.9. Record on your Lab Answer Sheet what you observe at three different times: immediately, about a minute later, and about 5 minutes later.

Applying What You Just Experienced
Review *clastic sedimentary rocks* in your textbook. Look especially for information about *bedding*.

Q16.10. What is the general term for the kind of bedding you watched form in the jar?

In areas of outburst flooding, repeated and extreme patterns of such bedding are called *rhythmites*. Rhythmites are common on small tributary streams that experience backflooding from the wall of water that occurs in an outburst flood. (Backflooding means that water rose so high in the main stream that it forced the flow of tributaries to reverse direction.)

Q16.11. In Figure 16.2, how many full cycles of flooding can you see?

Q16.12. Describe the shape of the individual particles you can see in the photo.

Q16.13. Using the 7-inch-high bottle in the photograph for scale, estimate the diameter of the largest particles washed in by the megaflood at this location.

Figure 16.2 Rhythmites near Starbuck, Washington. Note the 7-inch-high bottle for scale. *(Photo by E. K. Peters)*

Problems

Problem 16.1. Barringer Crater, Arizona. Arizona's "Meteorite Crater" is our planet's best-known meteorite impact feature. Geologists think that this great scar was created in seconds by an impact that occurred about 25,000 years ago. Study the aerial photo of the crater in Figure 16.3.

Q16.14. Using the width of the dirt roads visible in the photo, estimate the diameter of the crater, in kilometers.

Q16.15. Assume that the depth of the crater is about one-quarter of its diameter. Calculate the crater's volume. The volume of a crater may be approximated by this formula:

$$V_{crater} \approx (0.3) \, \pi \, (d/2)^2 \, D$$

where
$\pi \approx 3.14$
d = diameter of crater (in kilometers)
D = vertical depth of crater (in kilometers)

Figure 16.3 Barringer Crater, Arizona. *(John Sanford/Photo Researchers)*

Q16.16. Recall that energy is converted from one form to another but never lost. In your textbook or lecture notes, read about major processes that occur at the time of a meteor impact that help consume the meteor's kinetic energy (energy of motion). Name two such processes.

Problem 16.2. Peering at an Ancient Crater from the Space Shuttle. Figure 16.4 shows an area of Quebec as seen from the Space Shuttle.

Q16.17. Describe the shape of the lake you see.

Q16.18. Geologists are certain that this lake lies in the remnant of an old impact crater (age about 200 million years). The crater is 65 kilometers in diameter. Using the formula in Q16.15 and assuming a crater depth about one-quarter of the diameter, find the crater's volume in cubic kilometers.

Q16.19. What gradual process is currently contributing to the obliteration of this crater?

Q16.20. Consulting your textbook or lecture notes, explain why this crater structure looks like a ring rather than a bowl.

Problem 16.3. Weathering of Craters on Mars. It appears that, long ago, the surface of Mars featured flowing streams of water. Today it does not—Mars is basically a desert. The red planet does have a windy atmosphere,

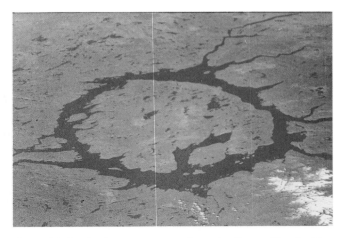

Figure 16.4 Ancient crater seen from the Space Shuttle. The crater is about 200 million years old and 65 kilometers in diameter. *(Courtesy of NASA)*

however, so some weathering continues on Mars. Figure 16.5 shows three Martian craters—**A, B,** and **C.** They display increasing degradation from weathering. Note which visual features of the craters appear or disap-pear with increasing weathering. Now examine crater **D** in Figure 16.6. It fits into the weathering sequence some-where among the other three.

Q16.21. Where does crater **D** in Figure 16.6 fit into the weathering sequence among craters **A, B,** and **C** shown in Figure 16.5?

Q16.22. Which of the four craters is likely to be the youngest?

Q16.23. Which kind of dating is your answer to the preceding question—relative or absolute?

Problem 16.4. Craters and Rivers on Mars. Figure 16.6 is another view of Mars. Study the craters shown. Also closely observe the channels of the braided streambed, now dry.

Q16.24. Are the craters generally older or younger than the stream channel?

Q16.25. Recall what you learned about relative dating of geologic events (Unit 7). What is the name of the principle you applied to answer the preceding question?

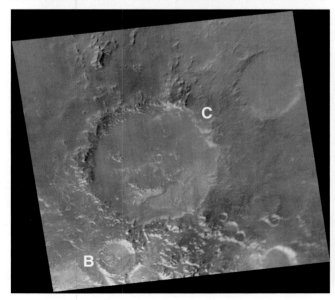

Figure 16.5 **Martian craters.** *(Courtesy of NASA/JPL/Malin Space Science Systems)*

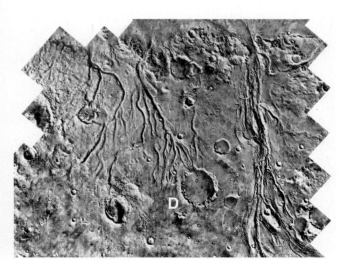

Figure 16.6 **Mangala Vallis Channels, now dry, left by a braided stream on Mars.** *(Courtesy of NASA)*

Problem 16.5. Megaripples in the Scablands of Washington State. Figure 16.7 is an aerial photo of megaripples. Use the buildings and line of rail cars to establish a rough scale.

Q16.26. Estimate (in feet) the average or typical distance from one crest of a megaripple to the next crest.

Q16.27. If you have waded barefoot in lakes or streams, you may have felt ripple marks in the mud or sand beneath your feet. Typical distances between crests of such ripples might be an inch. How many times larger than an inch is the distance you calculated for the megaripples in the photo? (This ratio suggests the enormous size of the megaflood in Washington State.)

Problem 16.6. Lower Moses Coulee. Figure 16.8 is an aerial photo showing the Lower Moses Coulee. Study the shadows at the center.

Q16.28. If the photo was taken in early afternoon to midafternoon, in what general direction must the camera have been pointed?

Q16.29. On the photo, outline the edge (top) of both sides of the coulee. Near the center and in the far distance, what is the general shape of the bottom of the coulee?

Figure 16.7   Megaripples in the Channeled Scablands. *(© John S. Shelton)*

Problem 16.7. Lower Grand Coulee. The largest of the coulees in the Scablands is appropriately called Grand Coulee. In 1940, part of the Columbia River valley nearby was blocked by Grand Coulee Dam to create a large reservoir. The kinetic energy of water released through the dam drives the turbines of Earth's third-largest hydroelectric power plant.

Figure 16.9 on page 243 is a topographic map that depicts part of the Grand Coulee area. Study the map, noting the many unusual features it displays.

Q16.30. Look closely at the feature named Dry Falls. It is so named because it looks a little like Niagara Falls—without any water! This enormous cliff cuts across the Grand Coulee. Using information on the map, what is the approximate elevation drop of Dry Falls?

Q16.31. At the base of Dry Falls are numerous lakes. According to the map, do Green Lake, Perch Lake, or Red Alkali Lake have outlets?

Q16.32. From the name Red Alkali Lake, and noting its lack of inlet and outlet streams, is its water likely to be drinkable? Why?

Q16.33. Lakes in the Scablands often are isolated from any drainage system. This is one reason that the area is described as having "chaotic drainage." Eastern Washington State is quite arid, so not all depressions are filled with water, as they are in wetter regions. How many closed depressions *without* lakes are shown on this map?

Q16.34. Notice the shape of Umatilla Rock. This rock is basalt. For some reason, this piece of basalt was especially resistant to being eroded away by the floods. On your Problem Answer Sheet, draw an overhead view (plan view or map view) showing the general shape of Umatilla Rock.

Figure 16.8   Lower Moses Coulee. *(© John S. Shelton)*

Q16.35. Study the shape of Umatilla Rock and the orientation of similar but smaller features elsewhere in the coulee. Name the compass direction *toward the source* of megafloods in this area.

Q16.36. The shape of coulees is one clue to the catastrophic origin of Scablands topography. On Figure 16.9 (the topo map), use a straightedge and pencil to draw a line drawn from point **A** to point **B.** Then, on Figure 16.10 on your Problem Answer Sheet, draw the topographic profile of the land.

Q16.37. Does the profile you've just drawn look like a normal river valley of moderate gradient?

Q16.38. The profile may remind you of roughly similar shapes found high in mountainous terrain. What geologic agent carves such areas in those places?

Q16.39. It is clear that glaciers did not carve the coulees of the Scablands. For one thing, there is no glacial till or moraine anywhere in the area. The coulee shapes were formed by a wall of water. Its energy ripped out basalt rock in long channels. In some places, the basalt was easier to remove than in others. Note points **C** and **D** on the map in Figure 16.9. Use a straightedge and pencil to draw a line from point **C** to point **D.** Then, on Figure 16.11 on the Problem Answer Sheet, draw the topographic profile of the land. Using a colored pencil, tint the rock that clearly was highly resistant to being ripped out of place by the catastrophic floodwaters.

Problem 16.8. Clark Fork Topographic Map. Remove the topographic map of Clark Fork, Idaho, from the back pocket of this book. Study the topography on the map.

Q16.40. Note the somewhat unusual valley in which Cascade Creek flows. Consider the part of the valley around the five buildings that lie just to the south of the unimproved road. In this area, how many streams run roughly parallel to Cascade Creek on the floor of the valley?

Q16.41. Follow the path of Cougar Creek as it flows down from Cougar Peak. When Cougar Creek reaches the valley in which Spring Creek runs, does it join Spring Creek?

Q16.42. This section of northern Idaho was the place where the torrential floodwaters of glacial Lake Missoula first broke through to the west. Describe the topographic evidence that the floodwaters were not contained within the current (modern) Clark Fork Valley.

Q16.43. In this area, floodwaters from Lake Missoula would have been approximately 1500 feet deep as they raced down the Clark Fork Valley. Knowing this, the locations of which map features may well have been underwater during the floods?

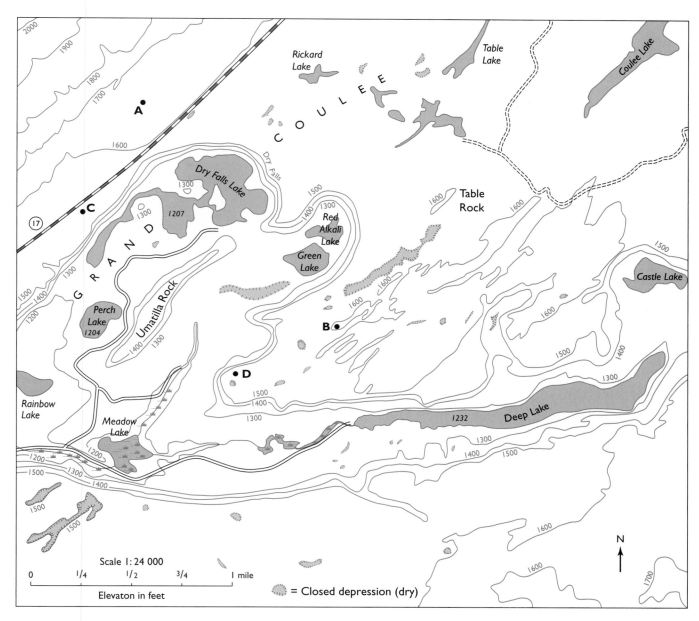

Figure 16.9   Topographic map of Lower Grand Coulee. *(1:24,000-scale Coulee City Quadrangle map. Adapted from U.S. Geological Survey.)*

Unit 16　Catastrophic Change

Last (Family) Name_____ First Name_____

Instructor's Name_____ Section_____ Date_____

Q16.1, 16.2.

Table 16.1	
Temperatures after pounding nails fully into wood	
Broad-headed nail	
Describe temperature of nail head	Describe temperature of hammer face
Small-headed finishing nail	
Describe temperature of nail head	Describe temperature of hammer face

Q16.3. _____

Q16.4. _____

Q16.5. _____

Explanation: _____

Q16.6. _____

_____

Q16.7. (circle one)　　faster　　slower　　can't tell

Q16.8. (circle one)　　more　　less　　can't tell

Q16.9. Immediately: _____

_____

_____

About a minute later: _____

_____

_____

About 5 minutes later: _____

_____

_____

Q16.10. _____ bedding

Q16.11. _____ cycles of flooding

Q16.12. _____

Q16.13. _____ diameter (give your units)

Unit 16 Catastrophic Change

Last (Family) Name_____ First Name_____

Instructor's Name_____ Section_____ Date_____

Q16.14. _____ kilometers

Q16.15. _____ cubic kilometers

Q16.16.

1. _____

2. _____

Q16.17. _____

Q16.18. _____ cubic kilometers

Q16.19. _____

Q16.20. _____

Q16.21.

_____ before **A**

_____ between **A** and **B**

_____ between **B** and **C**

_____ after **C**

Q16.22. (circle one)   **A**   **B**   **C**   **D**

Q16.23. (circle one)   relative   absolute

Q16.24. (circle one)   older   younger   can't tell

Q16.25. _____

Q16.26. _____ feet

Q16.27. _____ times larger

Q16.28. (circle one)   SE   SW   N

Q16.29. (circle one)   V-shaped   flat-bottomed

Q16.30. _____ feet

Q16.31. (circle one)   yes   no

Q16.32. (circle one)   yes   no

Explanation: _____

Q16.33. _____

Q16.34. General shape of Umatilla Rock:

Q16.35. (circle one)   NW   N   NE   E

Q16.36.

Figure 16.10 Topographic profile from points A to B on topographic map.

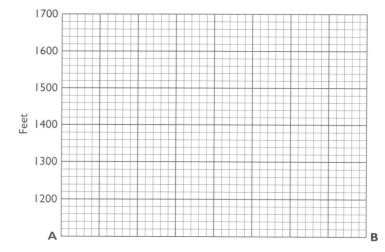

Q16.37. (circle one)   yes   no

Q16.38. _____

Q16.39.

Figure 16.11 Topographic profile from points C to D on topographic map.

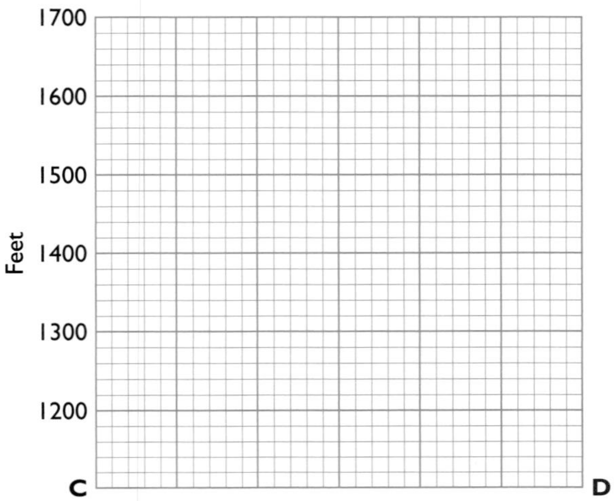

Q16.40. (circle one)   0   1   2   5

Q16.41. (circle one)   yes   no   not enough info

Q16.42. _____

_____

_____

Q16.43. (check all features whose locations probably were underwater)

_____ town of Clark Fork

_____ U. of I. Field Campus

_____ Lightning Creek

_____ Cascade Creek

_____ gravel pits between Cougar Creek and Spring Creek

_____ fish hatchery

_____ Antelope Lake

_____ falls on Johnson Creek

Geologists have long excelled at understanding the gradual changes we see acting on Earth's surface. These slow changes include glaciers creeping down mountains, the gradual shrinking of beaches along a shore, the expanding desert of North Africa—all of which may indicate climate change. These topics are the focus of this chapter.

Lab 17.1 Trends within Trends

The weather in any given year often seems "hotter than normal" or "colder than normal." This is not surprising: our concept of "normal" is based on an average, so the weather in any given year is likely to be different from the average. Nevertheless, when North America has an unusually cold winter, the tabloid headlines read "Return of Ice Age?" And in the same year, an unusually hot summer has them screaming "Global Hothouse!"

Such headlines are nonsense written to "sell product." But how about the previous dozen years? Have most of them been unusually cold or hot?

In this exercise, we alert you to the difficulties of accurately perceiving change over time. Then we sketch a way to think about global climate change.

Materials
• coin
• red and blue pencils

Procedure
1. Notice that Table 17.1 on the following page begins with a value of 35°F. Think of this value as representing the average nightly low temperature recorded over many years in the same month in a small town.

2. Flip your coin. If it comes up heads, *add* one degree to the 35 value and record it in the first blank in the table. If the coin comes up tails, *subtract* one degree from the 35 value and record it on the table.

3. Continue flipping, adding or subtracting *from your last value,* and recording each result until the table is complete.

Q17.1. Plot your data from the table on Figure 17.13 on your Lab Answer Sheet. Place a small dot on the graph for each datum. Then connect the dots with line segments.

4. Using a red pencil, shade the area(s) that lie *above* the 35° line but below the jagged line you constructed in Figure 17.13.

5. Using a blue pencil, shade the area(s) that lie *below* the 35° line but above the jagged line you constructed in the figure.

Applying What You Just Experienced
Your data set and your plot are similar to average weather records kept for the same month for one place over a 40-year time span. Indeed, if you looked at enough weather records across North America, you would find data highly similar to yours.

Q17.2. Because your data began at 35°, and because there was a 50–50 chance of change either upward or downward with each coin toss, you might expect your temperatures to stay near 35° throughout your record. In fact, what is the *highest* temperature you recorded?

Q17.3. What is the *lowest* temperature you recorded?

Q17.4. What is the *highest* temperature recorded by anyone in your class?

Q17.5. What is the *average highest* temperature in your class?

Q17.6. What is the *lowest* temperature recorded by anyone in your class?

Q17.7. What is the *average lowest* temperature in your class?

Q17.8. Which is more extreme: the *average* highest or lowest temperatures, or the *individual* highest or lowest temperatures? (Think about why this must be the case.)

Q17.9. Returning to your own data, how many major periods of upward trend do you see? (**Hint:** Use your own judgment about what constitutes a "major period.")

Q17.10. Do one or more of your upward-trending periods include at least some downward movement?

Table 17.1

Data set: Coin flips and temperature

Flip	°F	Flip	°F
1	35	21	
2		22	
3		23	
4		24	
5		25	
6		26	
7		27	
8		28	
9		29	
10		30	
11		31	
12		32	
13		33	
14		34	
15		35	
16		36	
17		37	
18		38	
19		39	
20		40	

Q17.11. How many major periods of downward movement do your data show?

Q17.12. Do one or more of your downward-trending periods contain at least some upward movement?

Q17.13. Can you see that reasonable people might disagree on the boundaries of upward and downward trends?

(If you take a course in statistics, you will discover how such issues are addressed quantitatively. Please consider taking such a course!)

Back to our point: scientists from many disciplines are investigating changes in global Earth temperatures. It's not possible to "take Earth's temperature" in any simple way. A colder-than-usual winter in California may occur at the same time as a warmer-than-usual winter in the high Arctic. So, how can scientists study fluctuations in global climate?

Scientists look at some variable other than temperature that they believe is related to global climate but less sensitive to local change. One such variable is the concentration of carbon dioxide (CO_2) in the atmosphere. Geologists and chemists who study ice cores from Antarctica and Greenland measure CO_2 concentrations in tiny air pockets trapped in the ice. Data of this sort from the past 220,000 years of the Pleistocene Epoch are shown in Figure 17.1A.

Q17.14. During the period shown, there are two relative high points. One is quite recent. The other occurred how many years ago?

Q17.15. From the older high point to about 20,000 years ago, there appears to be a strong downward trend in CO_2 concentration. From what you know of human history, can we reasonably credit this decrease to human activities? Why or why not?

Q17.16. For the past 20,000 years, what has Earth's atmosphere experienced in terms of CO_2 concentration?

Q17.17. Could the industrial burning of fossil fuels (a process that creates lots of CO_2) account for this broad trend? Why or why not?

Q17.18. It is possible that we humans are influencing very recent atmospheric CO_2 concentrations. Figure 17.1B shows data from the past two centuries. How would you describe the shape of the curve that the data appear to fit? (***Hint:*** Answer using a mathematical term if you can.)

Q17.19. Considering what you have learned about climate change, briefly explain what this CO_2 increase might mean for average global temperatures.

Q17.20. Even if world temperatures do increase, could your hometown still experience an extremely cold winter a decade from now?

Lab 17.2 Particle-Size Distributions Resulting from Gradual Processes

Geologists often study past Earth processes by considering the sediment they produce. Glaciers, rushing mountain streams, and tropical shores each produce very distinctive sediment.

Materials
• set of sieves

• mass balance

• two samples of sediment

Procedure
1. Study Table 17.2 on your Lab Answer Sheet. It is arranged in two parts, the upper part for sample **1** and the lower part for sample **2.** You will be measuring and weighing sample **1** and sample **2,** which consist of particles that were deposited by gradual geologic processes.

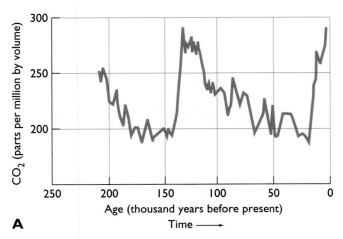

A

Time ⟶

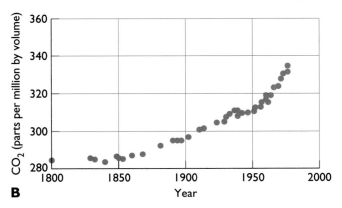

B

Figure 17.1 Carbon dioxide concentration in Earth's atmosphere. A. CO_2 (parts per million by volume, or ppmv) in air trapped in glacial ice during the past 220,000 years. **B.** Recent CO_2 concentration (ppmv) in Earth's atmosphere.

2. Weigh each sample.

Q17.21. Record their initial weights in the bottom row of each part of Table 17.2.

Q17.22. Record the sieve hole sizes (and thus the particle size that will be separated) in the second column of the table for both samples **1** and **2.**

3. Carefully sieve sample **1.** Weigh each size fraction (coarsest, next-finer, and so on).

Q17.23. Record your results in the third column of the upper part of the table.

Q17.24. Sum the weights of the size fractions in the next-to-last row of each part.

Q17.25. Calculate the percentage of total weight for each size fraction in the last column.

4. Compare these weights with the initial weight of the sample in the last row. If the initial weight agrees with the total weight within 5%, that is acceptable. If they disagree by more than 5%, reweigh everything and find your error.

Q17.26. Repeat step 3 and questions for sample **2.**

Q17.27. From the data in the table, construct histograms for sample **1** and sample **2** in Figure 17.14 on your Lab Answer Sheet. Label the Y axes on the figure.

Q17.28. From what you've learned, which gradual geologic process probably produced sample **1** and sample **2?** (*Hint:* Examine your samples for tiny bits of seashells.)

Q17.29. How would you describe the particles in each sediment?

Q17.30. How would you describe the sorting of each sample?

Applying What You Just Experienced

All gradual geologic processes require enormous periods of time to work, yet each process can produce its own distinctive sediment.

Q17.31. What are two gradual processes that can produce well-sorted sediment?

Q17.32. What very gradual geologic process is best known for producing sediment of extremely poor sorting?

Problems

Problem 17.1. Glacial Features Resulting from Gradual Processes (Palmyra, New York, Quadrangle). Open the large, folded map in the back of your manual. Find the map of Palmyra, New York. Note the many isolated hills shown on the map. Study their general shape and the direction in which they run.

Q17.33. What name do geologists use for such hills?

Q17.34. The gradual process that created these hills was glaciation. Which type of glaciation causes such hills?

Q17.35. In what general direction do the steeper slopes of these hills face?

Q17.36. Therefore, the glacial ice probably flowed from which direction?

Q17.37. Would you expect the hills to be made of solid bedrock or loose sediment?

Q17.38. In addition to the hills, name another topographic feature on the map that suggests the importance of glacial processes in shaping the land in this area.

Problem 17.2. Glacial Features in Iceland. Note the smooth surface of the rocks in Figure 17.2. This surface was gradually produced by glaciation.

Q17.39. What is the geologic term for the surface?

Q17.40. Ice alone could not have produced this smooth surface, because ice is much softer than rock. What helped produce the fine abrasion of this surface?

Q17.41. Knowing that the photo was taken in Iceland, and recalling what you have already learned about plate tectonics and rock types, what kind of rock has acquired this smooth surface?

Figure 17.2 Vista and rocks in Iceland.
(Photo by Pamela Judson-Rhodes)

Problem 17.3. Glacial Ice. Glaciers and rocks both deform under stress, and they do so in some similar ways. Study the deformation in the ice shown in Figure 17.3.

Q17.42. What part of a fold structure is visible in the left-hand part of the photo? (Note person for scale.)

Q17.43. What kind of stress does the glacial ice reflect?

Problem 17.4. Does Ice Flow Uphill? When an area with alpine glaciers experiences warmer winters than in previous years, the glaciers retreat up their valleys.

Figure 17.3 Ice in Fox Glacier, New Zealand. *(Photo by Sheldon Judson)*

Q17.44. Does this mean that the ice at the lower end of the glacier actually flows uphill? Explain.

Problem 17.5. Quantifying a Glacial Retreat. Mt. Rainier, a volcano near Seattle, has many glaciers on its slopes. One, the Nisqually Glacier, has been measured since 1857. Although the glacier has advanced during certain years, the record clearly shows it to be retreating overall—presumably in response to warmer climatic conditions in this locale. Study the record of the Nisqually Glacier's retreat in Figure 17.4 on the next page.

Take point **A** along the 1857 terminus as your starting point (zero) and measure the retreat of the glacier up its valley to the year 1870.

Q17.45. Record the distance (in meters) in Table 17.3 on your Problem Answer Sheet.

Now measure the distance up-valley from Point **A** to the 1885 terminus.

Q17.46. Record your result in the table.

Continue, each time starting from Point **A**, until you complete the table. (Follow the curve of the glacier where you need to.)

Q17.47. Now plot your data on Figure 17.15 on the Problem Answer Sheet to show time versus the cumulative retreat of the glacier.

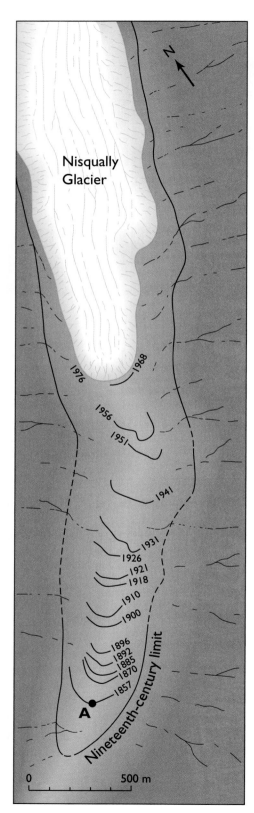

Figure 17.4 Generalized positions of terminus of Nisqually Glacier, 1857–1976. *(Courtesy of National Park Service and Sheldon Judson)*

Q17.48. Judging from your plot, what interval of years shows the fastest retreat of the glacier's terminus (the greatest slope on the figure)?

Q17.49. Considering the record as a whole, how many meters per year *on average* did the glacier retreat from 1857 to 1976?

Problem 17.6. Desert Features Resulting from Gradual Processes (Death Valley, California). Study the topographic map of Death Valley National Monument (Figure 17.5).

Q17.50. What is the general compass direction of the length of Death Valley?

Q17.51. To the east and west of the valley, ridges rise to significant elevations. What is the highest elevation shown east of the valley? West of the valley?

Q17.52. Which of the valley's walls, east or west, is steeper? Which has broad alluvial fans?

Q17.53. Local placenames often say a lot. List seven placenames hinting that this area is inhospitable—to put it mildly.

Q17.54. List four placenames indicating that this area has a mining history.

Q17.55. Most mining in this area recovers evaporite deposits. What is the geologic term for the body of water shown in Badwater Basin (near the center)?

Q17.56. Name two minerals likely to be concentrated in the Devil's Golf Course. Explain how they form in this environment.

Q17.57. What is the lowest elevation on the map?

Q17.58. What is the total topographic relief shown on the map (highest elevation minus lowest elevation)?

Q17.59. At what elevation does the map indicate a large area of vegetation?

Q17.60. What might account for the small patch of vegetation in the north-central part of the map?

Q17.61. If you visit Death Valley, what might be a wise time of year for a pleasant trip?

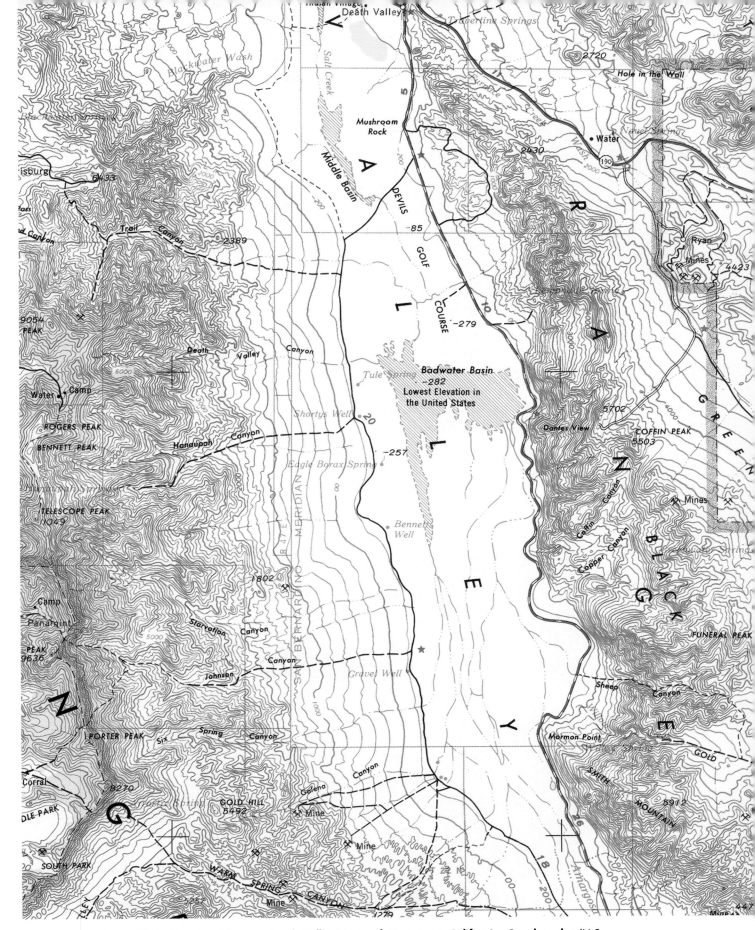

Figure 17.5 **Topographic map, Death Valley National Monument, California, Quadrangle.** *(U.S. Geological Survey)*

Figure 17.6 Broad apron of sediment on western side of Death Valley National Monument, California. *(Photo by L. Davis)*

Problem 17.7. Desert Processes and Streams. The photo in Figure 17.6 was taken in Death Valley.

Q17.62. What is the name for the broad apron of sediment that formed gradually and is shown in the photo?

Q17.63. Notice the shapes of the dry streambeds etched into the apron of sediment, and recall what you learned about streams earlier in this course. What word would a geologist use to describe the general pattern of these stream channels?

Q17.64. What does the channel pattern tell us about the loads carried by the streams when they are flowing?

Problem 17.8. Dune Types. In North Africa, dunes are common (Figure 17.7).

Q17.65. What is the geological term for the type of dune shown?

Q17.66. Do you see evidence that water sometimes flows in this desert? If so, what is your evidence?

Q17.67. What feature of the sand, in terms of sorting and roundness, would you expect to observe in these dunes?

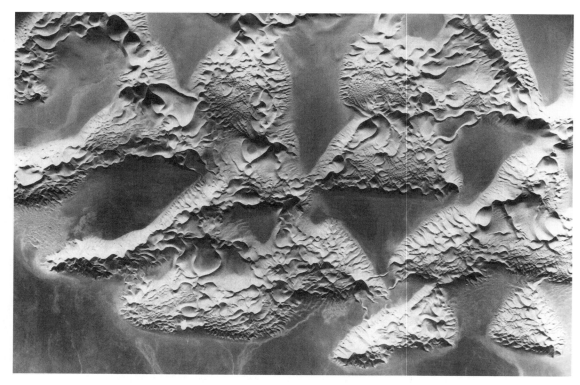

Figure 17.7 Aerial view of North African sand dunes. *(Photo by U.S. Army Air Corps)*

Figure 17.8 **Homes along the shore in La Jolla, California.** *(Photo by E. K. Peters)*

Problem 17.9. Shorelines and Houses. Study the shoreline in Figure 17.8.

Q17.68. Does the shoreline suggest net erosion or net deposition?

Q17.69. The million-dollar houses in the photo are in the San Diego suburb of La Jolla. Note the low white house above the bluff near the center. Is this house a good long-term investment or an insurance liability?

Problem 17.10. Types of Shorelines. Figure 17.9 shows a flat expanse of rock at low tide.

Q17.70. What do geologists call this flat feature?

Q17.71. What do geologists call the steep slope to the right?

Problem 17.11. Rising Sea Level. Some scientists worry that we are adding gasses to the atmosphere that could lead to global warming. Warming could melt the Antarctic ice cap, thus adding water to the oceans and raising sea level. We estimate the Antarctic ice cap to be about 2 kilometers thick, covering 9 million square kilometers. We also know the total surface area of the oceans is about 380 million square kilometers.

Q17.72. If glacial ice has a density of 0.92 metric tons per cubic meter and seawater has a density of nearly 1 metric ton per cubic meter, how many meters would sea level rise if all the ice in the Antarctic ice cap were to melt? (**Note:** Your answer will be a maximum estimate because, as sea level rises, total surface area of the oceans will also increase.)

Figure 17.9 **Low tide at Tomakins, New South Wales, Australia.** *(Photo by Sheldon Judson)*

Problem 17.12. Shoreline Features Resulting from Gradual Processes. Cape Hatteras National Seashore in North Carolina is famous for its beautiful barrier island beaches. A topographic map of this area is shown in Figure 17.10. The map is a portion of the Oregon Inlet Quadrangle, North Carolina. Study the map. It shows elevations on the island with the usual brown contour lines and water depths (in feet) with blue contour lines.

The tidal range at this location is about 2 feet. This means that, if you were standing at the water's edge at low tide, by the time the water depth peaks at high tide about 6 hours later, the water would be above your knees. Another 6 hours later, the water level would be back at your feet at low tide. The seashore elevation shown on the map is the average at high tide. The water depths shown are the average at low tide.

This map was first made in 1953. It was photorevised in 1983, thirty years later. Photorevision is done by flying over an area, photographing the ground, then updating the map with changed features observed in the photos. The changed features are printed in purple to make it easy to see what has changed since the original map was printed.

Q17.73. Examine the color of Highway 12, all the way from the top of the map to the bottom. What part of the highway was extended?

Q17.74. In the lower left quadrant of the map are three islands. During what span of years must they have formed?

Q17.75. Study the colors used to depict the shoreline along all of the islands on the map. Based on the colors you see, would you describe these islands as being stable or unstable in shape?

Q17.76. What is the highest elevation on the portion of the barrier island shown on the map?

Q17.77. On this map, seawater depth is shown in feet. (On many USGS maps, it is shown in meters.) What is the maximum water depth contour you see?

Q17.78. Every boat or ship floats at equilibrium, with so many feet of the vessel below the water's surface. The amount of the craft that is submerged is the *draft* of the vessel. Suppose you have a boat with a 4-foot draft. If you sailed it through Oregon Inlet at low tide and navigated along the dashed line marked "Boundary," what would happen?

Q17.79. Find the campground on the map. Was it constructed before or after 1953? If you wanted to enjoy the beautiful ocean beach, in which direction would you walk from the campground?

Q17.80. If you wanted to videotape the seabirds on Herring Shoal Island, how would you travel there? Explain your reasoning.

Q17.81. Find the benchmark in the southeastern quadrant of the map. Suppose that you are standing on this benchmark when Hurricane Fuerte strikes at high tide, creating a storm surge 9 feet high. Are your feet above flood level?

Q17.82. A longshore current flows along this coast. Study the map and decide in which direction it is flowing.

Q17.83. The Atlantic Ocean lies to the east of these barrier islands. Note the blue line labeled "Breakers" in the ocean. Why does this line curve instead of being straight?

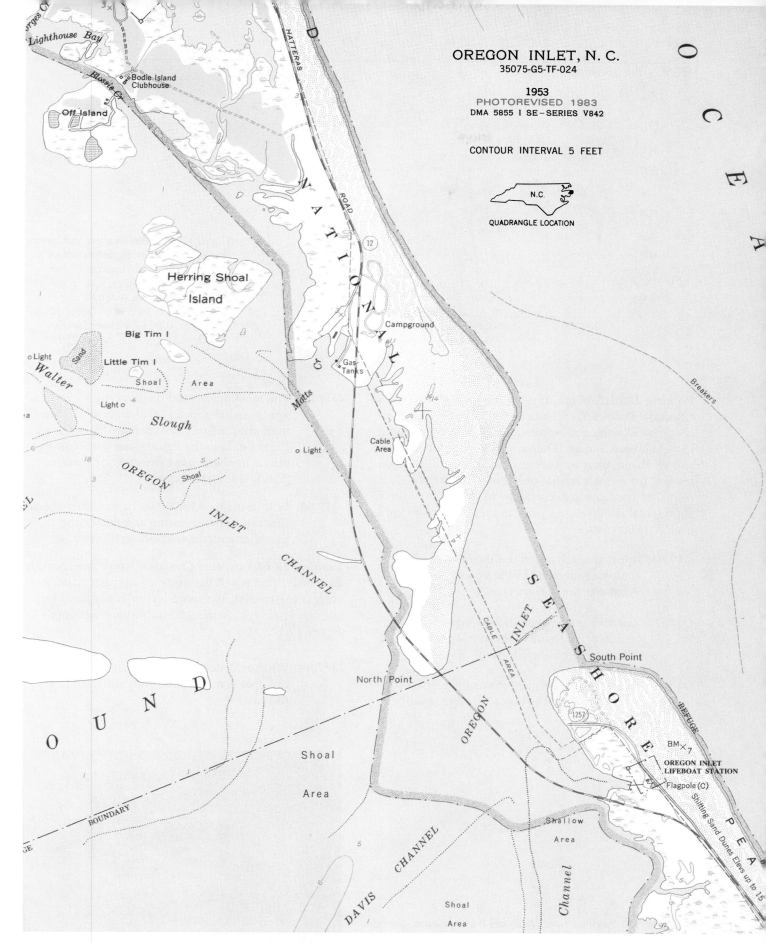

Figure 17.10 Topographic map, portion of the Oregon Inlet Quadrangle, North Carolina. (1:24,000 scale. Courtesy of *U.S. Geological Survey*)

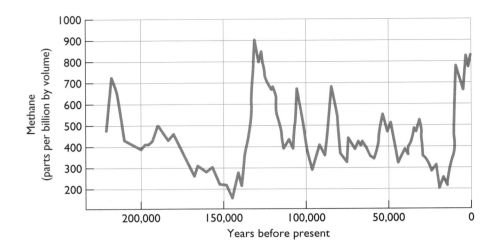

Figure 17.11 **Methane gas concentration in air pockets in continental ice sheets.** *(After J. Chappellaz, E. Brook, T. Blunier, and B. Malaize, CH$_4$ and δ^{18}O of O$_2$ records from Antarctic and Greenland Ice, Journal of Geophysical Research 102:C12, November 30, 1997)*

Problem 17.13. Global Climate Change Resulting from Gradual Processes. Scientists are studying current changes in Earth's atmosphere in light of concern about global climate change. Some gases, like carbon dioxide and methane, increase the amount of heat our planet retains from solar radiation. Study Figure 17.11, a record of methane concentration in the air pockets of continental ice sheets. As you can see, the data go back over 200,000 years.

Q17.84. Approximately when did the *lowest* concentration of methane in the atmosphere occur? What was the concentration?

Q17.85. Approximately when did the *highest* concentration of methane in the atmosphere occur? What was the concentration?

Q17.86. How many times greater was the highest concentration than the lowest concentration?

Q17.87. The most recent cold phase of the Pleistocene Epoch ended about 12,000 years ago. From then until today, global climate has warmed. Do the data in the figure show that atmospheric methane has increased or decreased during this warm period?

Q17.88. Within the past 12,000 years, have there been times of both increasing and decreasing atmospheric methane concentration?

Problem 17.14. Inferring Climate Change from Simple Evidence. The scientific study of climatic change is complex. However, the basic lesson is simple: Earth's climate is always changing! Study the photograph in Figure 17.12.

Q17.89. What evidence do you see that the climate in this area is now significantly different from the past?

Figure 17.12 **Petrified Forest National Park, Arizona.** *(Photo by Sheldon Judson)*

Geo Detectives at Work   THE CASE OF COAL ON THE ROCKS

Forensic geologists must be familiar with Earth's gradual processes and the way in which they leave their imprint on rocks and minerals. One such case involved an electrical generating plant in Massachusetts that burned coal. The grates within the plant's furnaces were being damaged by rocks that—somehow—got mixed in with the coal. Suspecting vandalism, the plant managers called in the police.

The first question was, Where were the rocks being added to the coal? Were they added at the coal mine hundreds of miles away, or to rail cars that carried the coal to Massachusetts, or at the plant itself?

A geologist studied the rocks and found that some showed glacial striations. This indicated they were probably part of the local glacial till or morraine, the unsorted mixture of glacial debris that ranges from boulders down to microscopic particles. In the area of the power plant, glacial till is overlain by soils that often contain glacially striated cobbles and boulders from the till. So, it was likely that the rocks were coming from near the plant, where the coal was stored in piles outdoors.

Further investigation disclosed that the heavy-equipment operator whose job was to scoop the coal from the stockpile was an alcoholic. When drinking, he got sloppy. He would scoop up more than coal, sometimes picking up the underlying soil when he neared the bottom of a coal pile. The fine soil particles did not cause a problem in the furnace, but the cobbles and boulders certainly did, breaking the furnace's grates. With this analysis, the plant managers were able to stop the problem.

Q17.90. What clue did the forensic geologist use to discover the origin of the rocks in the furnace?

Unit 17 Gradual Change

Last (Family) Name_____ First Name_____

Instructor's Name_____ Section_____ Date_____

Q17.1.

Figure 17.13 Data from coin flips.

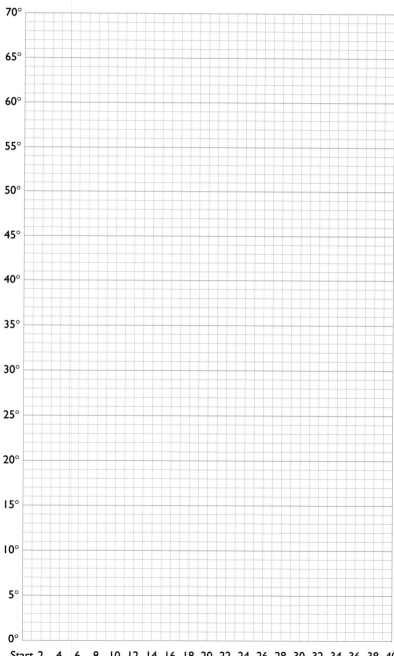

Q17.2. Your highest temperature: _____ °

Q17.3. Your lowest temperature: _____ °

Q17.4. Class highest temperature: _____ °

Q17.5. Class average of highest temperature: _____ °

Q17.6. Class lowest temperature: _____ °

Q17.7. Class average of lowest temperature: _____ °

Q17.8. (check one)

_____ Average highest or lowest temperatures were more extreme.

_____ Individual highest or lowest temperatures were more extreme.

Q17.9. _____ major upward periods

Q17.10. (circle one)   yes   no

Q17.11. _____ major downward periods

Q17.12. (circle one)   yes   no

Q17.13. (check one)

_____ Yes, reasonable people might disagree.

_____ No, it's quite black-and-white.

Q17.14. _____ years ago

Q17.15. (circle one)   yes   no

Explanation: _____

_____

Q17.16. (check one)

_____ little change

_____ generally downward, with some upward blips

_____ generally upward, with one downward blip

Q17.17. (circle one)   yes   no

Explanation: _____

_____

Q17.18. _____

Q17.19. _____

_____

Q17.20. (circle one)   yes   no

Q17.21, 17.22, 17.23, 17.24, 17.25, 17.26.

Table 17.2

Particle-size distributions
Sample 1

Particle size fraction	Particle diameter for this size fraction is greater than (show unit: millimeters or inches)	Weight of this size fraction (show unit: grams or ounces)	Percentage of total weight
Coarsest	>		%
Next-finer	>		%
Next-finer	>		%
Next-finer	>		%
Next-finer	>		%
	Sum the weights of all size fractions ⟹		100%
	Initial sample weight ⟹		

Sample 2

Particle size fraction	Particle diameter for this size fraction is greater than (show unit: millimeters or inches)	Weight of this size fraction (show unit: grams or ounces)	Percentage of total weight
Coarsest	>		%
Next-finer	>		%
Next-finer	>		%
Next-finer	>		%
Next-finer	>		%
	Sum the weights of all size fractions ⟹		100%
	Initial sample weight ⟹		

Q17.27.

Figure 17.14 **Histograms of Sample 1 and Sample 2 particle-size distribution.**

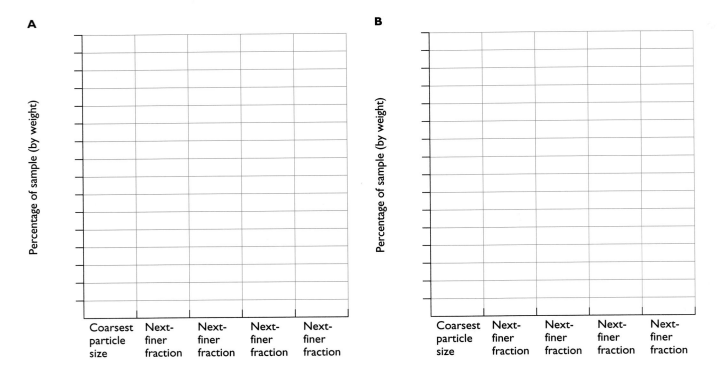

A

Percentage of sample (by weight)

Coarsest particle size | Next-finer fraction | Next-finer fraction | Next-finer fraction | Next-finer fraction

B

Percentage of sample (by weight)

Coarsest particle size | Next-finer fraction | Next-finer fraction | Next-finer fraction | Next-finer fraction

Q17.28. (check one for each sample)

Sample 1	Sample 2	
		wind blowing particles to form dunes
		glacial ice pushing sediment across an outwash plain
		ocean currents depositing sand on a beach

Q17.29. (check one for each sample)

Sample 1	Sample 2	
		rounded to well rounded
		angular to subangular

Q17.30. (check one for each sample)

Sample 1	Sample 2	
		very well sorted
		well sorted
		poorly sorted

Q17.31.

1. _____

2. _____

Q17.32. _____

Unit 17   Gradual Change

Last (Family) Name_____First Name_____

Instructor's Name_____Section_____Date_____

Q17.33. _____

Q17.34. (check one) _____ alpine glaciation _____ continental glaciation

Q17.35. (check one) _____ NNW _____ SSE

Q17.36. (check one) _____ NNW _____ SSE

Q17.37. (check one) _____ solid bedrock _____ loose sediment

Q17.38. _____

Q17.39. _____

Q17.40. _____

Q17.41. (circle one)   gneiss   slate or schist   granite   basalt

Q17.42. (circle one)   anticline   syncline   recumbent fold

Q17.43. (circle one)   extensional   compressional   can't tell

Q17.44. (circle one)   yes   no

Explanation: _____

_____

Q17.45, 17.46.

Table 17.3			
Retreat of the Nisqually Glacier			
Year	Cumulative meters	Year	Cumulative meters
1857	0 (starting value)	1921	
1870		1926	
1885		1931	
1892		1941	
1896		1951	
1900		1956	
1910		1968	
1918		1976	

Q17.47.

Figure 17.15   **Retreat of the Nisqually Glacier.**

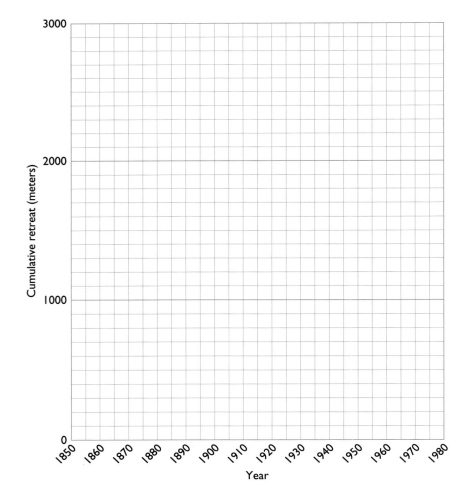

Q17.48. from _____ to _____

Q17.49. _____ meters per year

Q17.50. (circle one)   N–S   E–W

Q17.51.

east of the valley: _____ feet

west of the valley: _____ feet

Q17.52.

Steeper wall (circle one)   east   west

Wall with broad alluvial fans (circle one)   east   west

Q17.53.

1. _____

2. _____

3. _____

4. _____

5. _____

6. _____

7. _____

Q17.54.

1. _____

2. _____

3. _____

4. _____

Q17.55. _____

Q17.56. _____ and _____

Explanation: _____

_____

Q17.57. _____ feet

Q17.58. _____ feet

Q17.59. _____ feet

Q17.60. _____

Q17.61. (circle one)   July 4   January 1

Q17.62. _____

Q17.63. _____

Q17.64. (check one)

_____ high loads of sediment

_____ low loads of sediment

_____ no load of sediment

Q17.65. _____

Q17.66. (circle one)   yes   no   can't tell

Evidence: _____

Q17.67. (check one)

_____ poor sorting

_____ poor rounding

_____ good sorting and rounding

Q17.68. (check one)

_____ net deposition (more deposition than erosion)

_____ net erosion (more erosion than deposition)

_____ can't tell

Q17.69. (check one)

_____ good long-term investment

_____ insurance liability

Q17.70. _____

Q17.71. _____

Q17.72. _____ meters

Q17.73. _____

Q17.74. 19 _____ to 19 _____

Q17.75. (circle one)   stable   unstable

Q17.76. _____ feet

Q17.77. _____ feet

Q17.78. _____

_____

Q17.79. (circle one)   before 1953   after 1953

(circle one)   west   east

Q17.80. (circle one)   walk   swim   boat

Explanation: _____

_____

Q17.81. (circle one)   yes   no

Q17.82. (circle one)   SE to NW   NW to SE

Q17.83. _____

_____

Q17.84.

_____ years ago

_____ parts per billion by volume (ppbv)

Q17.85.

_____ years ago

_____ parts per billion by volume (ppbv)

Q17.86. _____ times greater

Q17.87. (circle one)   increased   decreased   can't tell

Q17.88. (circle one)   yes   no   can't tell

Q17.89. _____

_____

Q17.90. _____

Geology and Wealth

Part VIII

Economic Geology Unit 18

Our industrial society depends on raw materials from Earth. The production of oil, coal, natural gas, metals, and minerals is supported by the branch of Earth Science called **economic geology.** Economic geologists specialize in the exploration, discovery, and production of Earth's resources, employing their knowledge of all branches of geology in their work.

Economic geology combines geology with business. Economic geologists use their core knowledge of minerals, rocks, and rock structure to find natural deposits, but they must do so with an eye toward profitability. For example:

- A thick coal seam may look tempting, but problems of impurities or expensive transportation of the coal to market may render the coal economically useless.

- A large volume of natural gas may be found in a sandstone layer, but the layer may be so tightly cemented that the gas cannot be economically recovered from it.

- A valuable mineral deposit may extend beneath an area where mining cannot legally proceed.

Lab 18.1 Crystal Form of Gemstones in the Rough: Garnet

(**Note:** Review of your earlier work on minerals will help you with this lab.)

Gems fascinate us because of their color and the symmetry of their well-developed crystal faces. Both of these characteristics are determined by the chemical composition and atomic structure of these minerals.

Materials
- two samples (**A** and **B**) of the semiprecious gemstone garnet
- contact goniometer

Procedure
Using the contact goniometer as you did in Unit 3, "Minerals and Gems," measure the interfacial angles of the garnets.

Q18.1. What are the angles? Answer by completing Table 18.2 on the Lab Answer Sheet.

Q18.2. How many sides (faces) does each garnet display?

Applying What You Just Experienced
The shapes (crystal forms) into which garnets grow can be most intriguing. These shapes are constrained by some basic geometric principles. One such principle is that *regularly shaped solids with a particular number of sides will have a certain angle between the sides.* For example, a cube has six sides and the angle between any two adjacent sides is 90°. (Fill in the answers to the questions below on your Lab Answer Sheet.)

Q18.3. Garnets often grow into forms with twelve sides. What are these forms called?

Q18.4. A geologist can recognize the shape of a garnet even from a broken shard of the mineral because of what characteristic between the faces of the crystal?

Q18.5. Why might your two garnet samples have markedly different colors?

Lab 18.2 Crystal Form of Gemstones in the Rough: Diamond

People are so accustomed to seeing gemstones like diamonds after they have been shaped by lapidaries (gem cutters) that they sometimes confuse the work of Mother Nature with that of expert gem cutters. This exercise will help clear up such confusion.

Materials
- two small samples (**C** and **D**) of "rough" or uncut diamond (so small they are not worth pilfering!)
- hand lens for viewing the diamonds

Procedure
1. Examine one of the diamonds with the hand lens. Look carefully. It may help to rotate the sample slowly so that light strikes its sides from different angles.

2. Now do the same for the second diamond. Be patient and examine it carefully.

3. Refer to the information about crystal forms earlier in this book and in your textbook.

Q18.6. Describe the crystal form of the two samples.

Applying What You Just Experienced
Recall from your earlier study of minerals that diamond is not a silicate, like quartz. Instead it is made entirely of atoms of only one element, carbon. Diamond is a gemstone in part because it is the hardest mineral on Earth. The hardness of diamond is a direct result of the chemical bonds that hold the carbon atoms together.

Q18.7. Recalling what you learned earlier about minerals, what kind of chemical bond binds carbon atoms together?

The two diamonds that you examined both belong to one symmetry class, known as the isometric (equal sides) class. This fact reflects the symmetry of the crystal structure within diamond.

Q18.8. Were the shapes you saw in your two diamond samples created by Earth's natural processes or by humans?

Diamonds are cut to have definite, precise angles. The angles help to reflect light back and forth within the gem, increasing its reflectance and therefore its brilliance. (If anyone in your class is wearing a diamond, study its cut with a hand lens.)

Diamond is known for its strong refraction (ability to bend light). The degree of refraction depends on the wavelength of the light. When sunlight shines on a well-cut diamond, the diamond becomes a prism, creating tiny rainbows of color. These colors give a valuable diamond its "fire."

When gem cutters cut a rough diamond, they can choose from many different styles. A "brilliant" cut is one such style. It has many faces and is designed to catch light from almost any angle, making the diamond sparkle. Most diamonds for rings sold in this country are "brilliants" in the language of jewelers.

Lab 18.3 Four Common Hydrothermal Minerals and Their Environmental Impact

Gold, silver, copper, lead, and tin may all become concentrated in veins within host rocks. The veins form due to hydrothermal processes—the slow circulation of hot fluids within rocks. The most common minerals mined from veins are sulfides (not silicates, as you might expect). It will be helpful to review sulfide minerals in your textbook before starting this exercise.

Materials
- four samples of common hydrothermal minerals labeled **E, F, G, H**
- standard materials for testing hardness, streak, etc.

Procedure
1. Examine the minerals, paying particular attention to their luster. As you can see, samples **E, F,** and **G** each have metallic luster.

Q18.9. How would you describe the luster of sample **H?**

2. Examine the samples again and heft each one.

Q18.10. Which sample feels like it is the most dense?

Q18.11. Using the tools you've been given for the Mohs scale (glass, steel, etc.), determine the hardness range for each mineral. Complete Table 18.3 on the Lab Answer Sheet as you do your work.

Applying What You Just Experienced

You have examined four of the most common sulfide minerals found in ore deposits. Economic geologists recognize dozens of sulfide minerals in their daily work. Exploration geologists know these minerals as well as they know the alphabet.

Q18.12. Use Table 18.1, the identification key for economic minerals, on the following page, to identify samples **E, F, G,** and **H.** In Table 18.4 on your Lab Answer Sheet, name the four ore minerals and give their chemical compositions.

Q18.13. Some ore minerals are mined for their minor (trace) elements. For example, galena is sometimes an ore of silver. If this mineral is mined and processed only for its silver, what toxic element will be abundant in the mill tailings?

Lab 18.4 Economic Mineral Identification

In this lab, you will practice identifying mineral specimens of economic value.

Materials

- mineral specimens of economic value numbered 1 through 20.

Procedure

The minerals and metals that have economic value are diverse. It takes time and patience to identify them. In general, the identification key for economic minerals (Table 18.1) breaks down into these categories:

A. Precious metals, like gold and platinum
B. Gems—precious and semiprecious, like garnet or sapphire
C. Silica-rich minerals, like amethyst quartz
D. "Softies," like graphite and talc (can be scratched with your fingernail)
E. Magnetic minerals, like magnetite
F. Other economic minerals, like pyrite and galena

Q18.14. Use the key in Table 18.1 to identify each mineral. Use the tools your instructor has given you, such as a magnet and a piece of quartz, as well as the other tools you used to identify rock-forming minerals. Record your responses in Table 18.5 on the Lab Answer Sheet. State the gem name if any, and give at least one use for each mineral (in other words, a reason we mine it).

Problems

Ore samples vary widely in how much metal is present. The richness (or grade) of ore often is indicated in parts per million, or ppm. Thus, ore that contains 52 ppm gold has 52 parts of gold (by weight) for each million parts of ore—for example, 52 pounds of gold for each million pounds of ore. A metric ton—a million grams—of the ore would have 52 grams of gold within it. (*Note:* The price of gold and other precious metals is generally based on an old unit, the troy ounce, or "troy oz." There are 12 troy ounces in 1 pound.)

Problem 18.1. Grade and Value of an Ore Sample. A prospector samples a gold-bearing quartz vein and takes the samples to a lab. Together the samples weigh 27 kilograms. The lab reports that the ore has an average grade of 7 ppm gold—quite a high value.

Q18.15. How much gold is in the samples in milligrams? In troy ounces? Record your answers on the Problem Answer Sheet.

Q18.16. If gold is selling for US $303 per troy ounce, what is the value of the gold in the prospector's sample?

Problem 18.2. Profitability of a Silver Mine. A silver mine in northern Idaho produces 50,000 short tons of ore per day. (*Note:* One short ton = 2000 pounds.)

Q18.17. If the grade of the ore is 44.4 ppm silver, and if silver currently sells for US $5 per troy ounce, what is the gross value of the silver produced each day?

Problem 18.3. Lead from a Silver Mine. Lead is a common by-product of silver mining.

Q18.18. If the mine in Problem 18.2 produces one-quarter of a pound of lead for each troy ounce of silver produced, how much lead (in pounds) is generated each day?

Q18.19. Approximately how many times greater is the concentration of lead than of silver in the Idaho mine?

Table 18.1

Identification key for economic minerals

Step 1 Study the sample carefully. If a sample contains more than one mineral, be sure you understand which one is of economic interest (check with your instructor if in doubt).

Step 2 Read the general descriptions in the following list. Decide which one best fits your sample. Then go to the specified section of the identification key.

• Sample contains small (even tiny) metallic specks surrounded by other minerals that are noneconomic (often pyroxene or quartz). Or sample appears to be tiny flakes or dust of metallic luster. Go to **A. Precious Metals.**

• Sample has a vitreous (glassy) luster and is very hard—it easily scratches your piece of quartz. Go to **B. Gems—Precious and Semiprecious.**

• Sample sometimes scratches quartz, but not easily—requires some effort. Go to **C. Silica-Rich Minerals.**

• Sample can be scratched with your fingernail. Go to **D. "Softies."**

• Sample is magnetic to some degree. Go to **E. Magnetic Minerals.**

• Sample does not fit any of the above descriptions well. Go to **F. Other Economic Minerals.**

A. Precious Metals

Gold: Rich yellow to light yellow; soft; metallic luster; often surrounded by mafic minerals such as pyroxene or suspended in massive quartz. May also be small, isolated flakes, "dust," or nuggets, pale to rich yellow.

Platinum: White to silver; soft; metallic luster; generally suspended in mafic minerals such as pyroxene. May also be small, isolated flakes or nuggets, white to silver in color. (See **F** for notes on native silver.)

B. Gems—Precious and Semiprecious

Diamond: Small cubes, octahedra, or broken shards. Clear or pastel tints. Highest degree of hardness (10) and luster (adamantine).

Corundum: Barrel-shaped, six-sided crystals or small grains; may be light colored to black; no cleavage; vitreous luster; hardness = 9. Gray to black variety is *emery.* Red variety is *ruby.* Blue and all other tints are known as *sapphire.*

Beryl: Often hexagonal crystals, but can be massive; many colors, all with vitreous luster; poor cleavage, conchoidal fracture; hardness = 8. Green variety is *emerald.* Blue-green variety is *aquamarine.*

Topaz: Clear or pale tints; vitreous luster; hardness = 8; stubby crystals common; perfect cleavage in one direction.

Garnet: Color variable; 12-, 24-, 36-sided crystals; no cleavage; vitreous luster; hardness = 6–7.5. Deep yellow-red variety is *pyrope;* violet-red variety is *almandine;* brownish to black variety is *spessartine;* pale tints usually *grossular* or *andradite.*

C. Silica-Rich Minerals

Consider the following possibilities. All lack cleavage but may show conchoidal fracture.

Vitreous luster

Amethyst: purple, *citrine:* yellow-brown, *rose quartz:* pink, *"rock quartz":* clear, *smoky quartz:* clear but smoky, dark

Duller than vitreous luster

> *Jasper:* rust-red color, *tiger eye:* light and dark yellow bands

D. "Softies"

Graphite: Dark gray-black; submetallic luster; looks like the "lead" in a pencil.

Gypsum: Clear or white; sometimes transparent; one perfect cleavage; may be earthy or occur in masses. Can be scratched easily with a fingernail.

Talc: Hardness = 1; pale colors; pearly luster; massive; soapy feel, like talcum powder.

Clays: Masses of microscopic crystals you cannot see with your unaided eye; dull luster; white, reddish brown to black; earthy odor, especially when dampened by breath.

Bauxite: Usually massive, sometimes small brown spheres in a lighter matrix; dull luster; usually a little harder than clays and lacks an earthy odor.

Halite: Can be scratched by a fingernail; hardness = 2–2.5; usually clear or white; perfect cubic cleavage; vitreous luster; tastes like table salt; highly soluble.

Sylvite: Can be scratched with a fingernail; hardness = 2; colorless or pale hues; perfect cubic cleavage; vitreous luster; taste is metallic; highly soluble.

E. Magnetic Minerals

Strongly magnetic

Magnetite: Often granular; may be octahedral; noticeably dense; black with metallic luster; hardness = 5–6.5; density = 5.2 grams/cubic centimeter.

Weakly magnetic

Pyrrhotite: Often massive or granular; bronzy black-brown color; metallic luster; hardness = 4 (or greater, due to impurities).

F. Other Economic Minerals

Malachite: Rich green; often occurs with azurite; effervesces (fizzes) in weak acid.

Azurite: Deep azure blue; often laced with malachite; effervesces in weak acid.

Pyrite: Cubes with striated faces; rarely octahedral; massive; light yellow; metallic luster; hardness = 6–6.5; streak is greenish black.

Chalcopyrite: Richer yellow than pyrite; usually massive; often with iridescent tarnish; metallic luster; hardness = 3.5–4; black streak.

Galena: Cubic or octahedral crystals; massive; gray, highly metallic luster; hardness between 2 and 3; noticeably dense; good cubic cleavage.

Sphalerite: Tetrahedral crystals or massive; submetallic to resinous luster; often brown but may be yellow, green, or black; hardness = 3.5–4; whitish streak; perfect dodecahedral cleavage.

Native copper: Distorted crystals or masses; copper color and metallic luster but may be highly tarnished; hardness = 2.5–3; noticeably dense.

Native silver: Crystals rare, but wires and dendritic masses more common; silver color and metallic luster but usually tarnished, sometimes to a dull black; hardness = 2.5–3; noticeably dense.

Fluorite: Cubes or octahedrons with twins; many colors, often pastels; hardness = 4; perfect octahedral cleavage; no reaction to weak acid; highly vitreous luster.

Halite: Cubic crystals or massive; good cubic cleavage; colorless to white to reddish; hardness = 2.5; highly soluble in water; salty taste.

Staurolite: Crystals grow in crosses; dark brown with vitreous luster; hardness = 7–7.5

Problem 18.4. The Lowest-Grade Ore on Earth: Diamond Ore. Diamonds cost more by weight than any other natural substance on Earth. Thus, companies and individuals are willing to mine lower grades of ore for diamonds than they would for other substances. Diamonds are sold by tiny units of weight called **carats;** 5 carats equal 1 gram. A typical diamond mine may process 100 metric tons of ore for each 10 carats of diamond.

Q18.20. What is the grade of such ore in terms of parts per million (ppm)?

Problem 18.5. Historical Mining Techniques. Study the two photos from Montana and Colorado, taken in 1871 and 1875, respectively, in Figure 18.1.

Q18.21. What type of ore deposit are the men mining? (**Hint:** You cannot determine what metal the men are mining, but the general type of ore deposit is clear.)

Q18.22. The lone man laboring in Montana in 1871 is using what tool to separate worthless material from the valuable mineral he is seeking?

Q18.23. The three men shown in Colorado in 1875 have built a device to help them separate worthless material from metal particles. This device is known as what kind of box?

Problem 18.6. Interpreting a Cross Section Beneath Two Mineral Claims. Study the cross section in Figure 18.2. The person shown is jubilant because she just realized that a gold-bearing quartz vein is at her feet. Notice that from her viewpoint at the surface she cannot know about the geology hidden beneath her. But you can see the cross section, so you can answer the following questions.

Q18.24. What type of fault has offset the rocks of this area?

Q18.25. Did the quartz vein form before, after, or at the same time the fault occurred?

Q18.26. Which mineral claim, **X** or **Y,** has more gold ore at the surface?

Q18.27. Which mineral claim has more gold ore when all depths are combined?

A

B

Figure 18.1 A. Madison County, Montana, 1871. B. Borens Gulch, La Plata County, Colorado, 1875. *(Both photos by W. H. Jackson, courtesy of U.S. Geological Survey)*

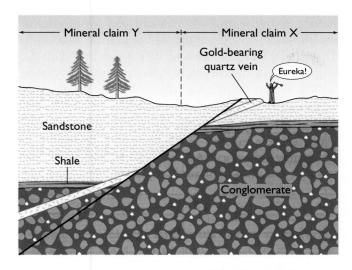

Figure 18.2 **Cross section beneath two mineral claims.**

Q18.28. What type of fault has offset the rocks of this area?

Q18.29. Is vein **X** a continuation of vein **Y** or vein **Z**?

Q18.30. Seeing only what a geologist standing on the surface could see, and assuming that the material in veins **X** and **Z** is similar, is the geologist likely to infer that vein **Y** is there, or could vein **Y** go unsuspected indefinitely?

Problem 18.7. Interpreting a Cross Section with Faults.
Study the cross section in Figure 18.3.

Q18.31. Veins **Y** and **Z** are nearly parallel. This is because veins typically occupy what type of structure?

Problem 18.8. Interpreting a Cross Section with Coal.
Imagine that you are a geologist working for a coal company in China. You map a flat region, compiling your field information to create the geologic map in Figure 18.4.

Q18.32. What is the full name of the geologic structure you have mapped?

Q18.33. Another company has claimed the coal visible at the surface. But four mineral claims are still available in the area. They are labeled **Q, R, S,** and **T** on the map. Which claim is likely to contain the most coal?

Q18.34. Will the others contain any coal?

Q18.35. Is the fault shown younger or older than the coal bed?

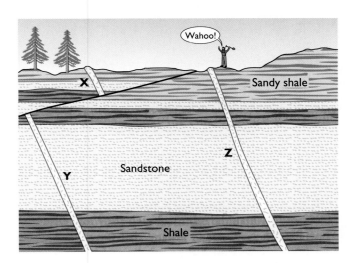

Figure 18.3 **Cross section.**

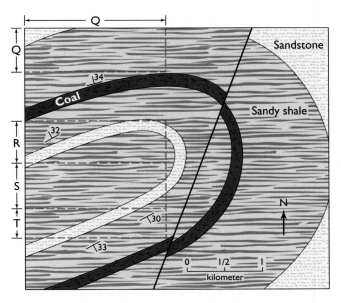

Figure 18.4 **Geologic map of coal claims in China.**

Problem 18.9. Climax Molybdenum Ore Deposit in Colorado. Molybdenum is a valuable metal. It can be alloyed with iron to create strong steel ("moly steel"). Molybdenum deposits often form as large "blankets" or "caps" of altered rock. They form due to the action of hydrothermal fluids above granite.

The most common molybdenum ore is molybdenum sulfide, known as *molybdenite.* It is a slippery substance that makes a good "dry" lubricant. Molybdenum ore is generally capped by a zone of tungsten-rich rocks. Sometimes granite stocks move upward in the same area in several different pulses, leading to overlapping caps of molybdenum ore. Figure 18.5 shows an idealized version of this concept in cross section.

In the real world, the situation can be quite messy and hard to interpret. At the huge Climax molybdenum mine in Colorado, geologists faced a situation like that shown in Figure 18.6. As you can see, two bodies of molybdenum ore exist at the surface of the Climax area. The mining company immediately began to exploit the larger one, which is exposed at lower elevation. Mining geologists also investigated the smaller and higher ore body, known as the Ceresco ore. By studying core samples, geologists constructed the cross section you see in the figure. The Mosquito Fault is a normal fault, with a vertical offset of rocks along the fault of 9200 feet (1.7 miles—quite an offset!).

Obviously, erosion has removed a lot of the Ceresco ore. But, if more of the Ceresco body is to be found, where would it be? Sketch your answer directly on the figure.

Q18.36. Which labeled point is nearest where you predict the rest of the Ceresco ore would be?

A. One stock

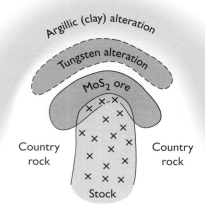

B. Repeated pulses of magma

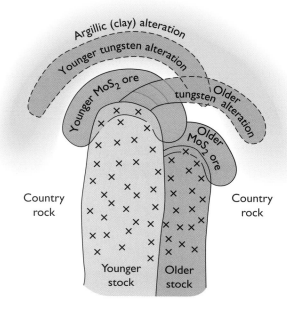

Figure 18.5 **Idealized cross sections of molybdenum ore.**

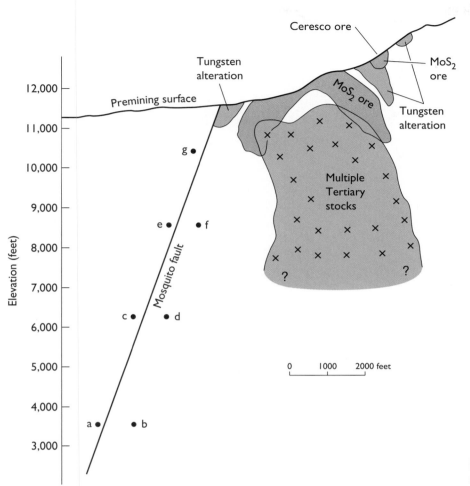

Figure 18.6   Slightly simplified cross section of the Climax Mine in Colorado.

Unit 18   Economic Geology

Last (Family) Name_____First Name_____

Instructor's Name_____Section_____Date_____

Q18.1.

Table 18.2

Interfacial angles of the garnet samples

Measurement	Garnet A angles	Garnet B angles
1		
2		
3		
4		
5		
6		

Q18.2.

Garnet **A:** _____ sides

Garnet **B:** _____ sides

Q18.3. _____

Q18.4. _____

Q18.5. (check one)

_____ different crystal structure

_____ different chemical composition

_____ ionic instead of covalent chemical bonds

Q18.6.

Sample **C:** _____

Sample **D:** _____

Q18.7. _____ bond

Q18.8. (check one)

_____ Earth's natural processes

_____ humans

Q18.9. _____

Q18.10. (circle one)   E   F   G   H

Q18.11.

Table 18.3

Hardness of samples **E, F, G,** and **H**

Sample	Harder than	Softer than	Therefore, Mohs scale hardness of sample is about
E			
F			
G			
H			

Q18.12.

Table 18.4

Ore sample names and compositions

Ore sample	Name	Chemical composition
E		
F		
G		
H		

Q18.13. _____

Q18.14.

Table 18.5

Mineral identification and use

Sample	Relevant features you see (luster, hardness, color, etc.)	Mineral name (and gem name if appropriate)	Use (reason for mining)
1			
2			
3			
4			
5			
6			
7			
8			
9			
10			
11			
12			
13			
14			
15			
16			
17			
18			
19			
20			

Unit 18 Economic Geology

Last (Family) Name_____First Name_____

Instructor's Name_____Section_____Date_____

Q18.15. _____ milligrams; _____ troy ounces

Q18.16. US $ _____

Q18.17. US $ _____

Q18.18. _____ pounds

Q18.19. _____ times greater

Q18.20. _____ ppm

Q18.21. _____

Q18.22. _____

Q18.23. _____ box

Q18.24. (circle one)   normal   reverse   strike-slip

Q18.25. (circle one)   before   after   same time

Q18.26. (circle one)   **X**   **Y**

Q18.27. (circle one)   **X**   **Y**

Q18.28. (circle one)   normal   reverse   strike-slip

Q18.29. (circle one)   **Y**   **Z**   can't tell

Q18.30. (check one)

_____ Vein **Y** might remain unsuspected and undetected for a long time.

_____ Vein **Y** is obvious from surface information alone.

Q18.31. _____

Q18.32. _____

Q18.33. (circle one)   **Q**   **R**   **S**   **T**

Q18.34. (circle one)   yes   no

Q18.35. (circle one)   younger   older   can't tell

Q18.36. (circle one)   **a**   **b**   **c**   **d**   **e**   **f**   **g**

INDEX

Note: Page numbers in *italics* indicate illustrations; those followed by t indicate tables.

LENGTH

12 inches = 1 foot
3 feet = 1 yard
5280 feet = 1 mile
6 feet = 1 fathom
0.394 inches = 1 centimeter
100 centimeters = 1 meter
39.4 inches = 1 meter
1000 meters = 1 kilometer
1.609 kilometers = 1 mile

VOLUME

2 cups = 1 pint
2 pints = 1 quart
4 quarts = 1 gallon
0.264 gallon = 1 liter
3.79 liters = 1 gallon
1 cubic centimeter = 1 milliliter
1000 milliliters = 1 liter

WEIGHT/MASS

16 ounce (oz) = 1 pound (lb)
12 troy ounces = 1 pound (lb)
28.4 grams = 1 ounce
1000 grams = 1 kilogram
2.205 pounds = 1 kilogram
2000 pounds = 1 (English) ton
1000 kilograms = 1 metric tonne

SPECIAL EQUIVALENTS

1 pint of water weighs very close to 1 pound
1 cubic centimeter of water weighs 1 gram
1 bar of pressure is very close to 1 atmosphere of pressure

TOPOGRAPHIC MAP SYMBOLS

VARIATIONS WILL BE FOUND ON OLDER MAPS

Primary highway, hard surface.............................

Secondary highway, hard surface ....

Light-duty road, hard or improved surface..................

Unimproved road ....

Road under construction, alinement known..................

Proposed road...

Dual highway, dividing strip 25 feet or less.................

Dual highway, dividing strip exceeding 25 feet ....

Trail ....

Railroad: single track and multiple track ....

Railroads in juxtaposition ....

Narrow gage: single track and multiple track ....

Railroad in street and carline ....

Bridge: road and railroad................................

Drawbridge: road and railroad ....

Footbridge ....

Tunnel: road and railroad ....

Overpass and underpass ....

Small masonry or concrete dam ....

Dam with lock ....

Dam with road..

Canal with lock ....

Buildings (dwelling, place of employment, etc.) ....

School, church, and cemetery ....

Buildings (barn, warehouse, etc.) ....

Power transmission line with located metal tower ....

Telephone line, pipeline, etc. (labeled as to type)..........

Wells other than water (labeled as to type).............. o Oil .... .o Gas

Tanks: oil, water, etc. (labeled only if water) .... ● ● ● ⊘ Water

Located or landmark object; windmill .... o .... ⚒

Open pit, mine, or quarry; prospect...................... ⚒ x

Shaft and tunnel entrance .... ◪ .... Y

Horizontal and vertical control station:

    Tablet, spirit level elevation .... BM △ 5653

    Other recoverable mark, spirit level elevation .... △ 5455

Horizontal control station: tablet, vertical angle elevation.. VABM △ 9519

    Any recoverable mark, vertical angle or checked .....

    elevation .... △ 3775

Vertical control station: tablet, spirit level elevation....... BM × 957

    Other recoverable mark, spirit level elevation .... × 954

Spot elevation .... × 7369   × 7369

Water elevation....................................... 670   670

Courtesy of USGS.

Boundaries: National ....

    State ....

    Country, parish, municipio ....

    Civil township, precinct, town, barrio ....

    Incorporated city, village, town, hamlet ....

    Reservation, National or State ....

    Small park, cemetery, airport, etc. ....

    Land grant ....

Township or range line, United States land survey ....

Township or range line, approximate location ....

Section line, United States land survey ....

Section line, approximate location ....

Township line, not United States land survey ....

Section line, not United States land survey ....

Found corner: section and closing ....

Boundary monument: land grant and other................

Fence or field line ....

Index contour ....   Intermediate contour ....

Supplementary contour.. .   Depression contours ....

Fill ....   Cut

Levee...................   Levee with road.........

Mine dump ....   Wash ....

Tailings ....   Tailings pond ....

Shifting sand or dunes....   Intricate surface........

Sand area.............   Gravel beach ....

Perennial streams ....   Intermittent streams .....

Elevated aqueduct ....   Aqueduct tunnel ....

Water well and spring....   Glacier..............

Small rapids ....   Small falls ....

Large rapids............   Large falls ....

Intermittent lake ....   Dry lake bed ....

Foreshore flat ....   Rock or coral reef......

Sounding, depth curve ... 10   Piling or dolphin ......

Exposed wreck ....   Sunken wreck ....

Rock, bare or awash; dangerous to navigation ....

Marsh (swamp) ....   Submerged marsh ....

Wooded marsh ....   Mangrove

Woods or brushwood....   Orchard.............

Vineyard ....   Scrub ....

Land subject to   Urban area ....
controlled inundation ....